NEUROSCIENCE
INTELLIGENCE
UNIT 7

Aβ Metabolism
and Alzheimer's Disease

Takaomi Comings Saido

Laboratory for Proteolytic Neuroscience
RIKEN Brain Science Institute
Wako-shi, Saitama, Japan

CRC Press
Taylor & Francis Group
Boca Raton London New York

CRC Press is an imprint of the
Taylor & Francis Group, an **informa** business

Aβ Metabolism and Alzheimer's Disease

Neuroscience Intelligence Unit 7

Eurekah.com
Landes Bioscience
Designed by Celeste Carlton

CRC Press
Taylor & Francis Group
6000 Broken Sound Parkway NW, Suite 300
Boca Raton, FL 33487-2742

First issued in hardback 2019

© 2003 by Taylor & Francis Group, LLC
CRC Press is an imprint of Taylor & Francis Group, an Informa business

No claim to original U.S. Government works

ISBN-13: 978-1-58706-230-8 (hbk)

While the authors, editors and publisher believe that drug selection and dosage and the specifications and usage of equipment and devices, as set forth in this book, are in accord with current recommendations and practice at the time of publication, they make no warranty, expressed or implied, with respect to material described in this book. In view of the ongoing research, equipment development, changes in governmental regulations and the rapid accumulation of information relating to the biomedical sciences, the reader is urged to carefully review and evaluate the information provided herein.

Library of Congress Cataloging-in-Publication Data

CIP applied for but not received at time of publication.

Visit the Taylor & Francis Web site at
http://www.taylorandfrancis.com

and the CRC Press Web site at
http://www.crcpress.com

Dedication

To the late George F. Comings (1931-2002)*
Rotarian
and
My American Father,
Husband of Dorothy J. Comings
and
Father of Stephen & Benjamin Comings

* The Comings family gave me an international outlook on life when I lived with them as a high school exchange student in Corinth, New York, U.S.A., in 1976-1977.

CONTENTS

EDITOR

Takaomi Comings Saido
Laboratory for Proteolytic Neuroscience
RIKEN Brain Science Institute
Wako-shi, Saitama, Japan
email: saido@brain.riken.go.jp
Chapters 1, 6

CONTRIBUTORS

Masashi Asai
Department of Biochemistry
Faculty of Pharmaceutical Sciences
Tokyo University of Science
Shinjuku-ku, Tokyo, Japan
Chapter 3

Kelly R. Bales
Neuroscience Discovery Research
Eli Lilly and Co., Lilly Research
  Laboratories
Indianapolis, Indiana, U.S.A.
Chapter 11

Martin Citron
Department of Neuroscience
Amgen Inc.
Thousand Oaks, California, U.S.A.
email: mcitron@amgen.com
Chapter 2

Ronald B. DeMattos
Neuroscience Discovery Research
Eli Lilly and Co., Lilly Research
  Laboratories
Indianapolis, Indiana, U.S.A.
email:
  demattos_ronald_bradley@lilly.com
Chapter 11

Christopher B. Eckman
Mayo Clinic Jacksonville
Jacksonville, Florida, U.S.A.
email: Eckman@mayo.edu
Chapter 7

Elizabeth A. Eckman
Mayo Clinic Jacksonville
Jacksonville, Florida, U.S.A.
email: Eckman@mayo.edu
Chapter 7

Blas Frangione
Departments of Pathology
  and Psychiatry
New York University Medical Center
New York, New York, U.S.A.
Chapter 10

Todd E. Golde
Department of Neuroscience
Mayo Clinic Jacksonville
Jacksonville, Florida, U.S.A.
email: tgolde@mayo.edu
Chapter 8

Dieter Hartmann
Department for Human Genetics
K.U. Leuven
Flanders Interuniversity Institute
  for Biotechnology
Leuven, Belgium
email:
  dieter.hartmann@med.kuleuven.ac.be
Chapter 5

Chinatsu Hattori
Department of Life Sciences
Graduate School of Arts and Sciences
University of Tokyo
Meguro-ku, Tokyo, Japan
Chapter 3

David M. Holtzman
Department of Neurology
Alzheimer's Disease Research Center
Washington University School
  of Medicine
St. Louis, Missouri, U.S.A.
email: holtzman@neuro.wustl.edu
Chapter 11

Nika Hotoda
Department of Life Sciences
Graduate School of Arts and Sciences
University of Tokyo
Meguro-ku, Tokyo, Japan
Chapter 3

Melitza Iglesias
Department of Neurology
Center for Neurologic Diseases
Brigham and Women's Hospital
Harvard Medical School
Boston, Massachusetts, U.S.A.
Chapter 12

Yasuo Ihara
Department of Neuropathology
Faculty of Medicine
University of Tokyo
Bunkyo-ku, Tokyo, Japan
Chapter 9

Shoichi Ishiura
Department of Life Sciences
Graduate School of Arts and Sciences
University of Tokyo
Meguro-ku, Tokyo, Japan
email: cishiura@mail.ecc.u-tokyo.ac.jp
Chapter 3

Cynthia A. Lemere
Department of Neurology
Center for Neurologic Diseases
Brigham and Women's Hospital
Harvard Medical School
Boston, Massachusetts, U.S.A.
email: lemere@cnd.bwh.harvard.edu
Chapter 12

Jodi F. Leverone
Department of Neurology
Center for Neurologic Diseases
Brigham and Women's Hospital
Harvard Medical School
Boston, Massachusetts, U.S.A.
Chapter 12

Chica Mori
Department of Neurology
Center for Neurologic Diseases
Brigham and Women's Hospital
Harvard Medical School
Boston, Massachusetts, U.S.A.
Chapter 12

Maho Morishima-Kawashima
Department of Neuropathology
Faculty of Medicine
University of Tokyo
Bunkyo-ku, Tokyo, Japan
email: maho@m.u.-tokyo.ac.jp
Chapter 9

M. Paul Murphy
Department of Neuroscience
Mayo Clinic Jacksonville
Jacksonville, Florida, U.S.A.
Chapter 8

Hiroyuki Nakahara
Laboratory for Mathematical
    Neuroscience
RIKEN Brain Science Institute
Wako-shi, Saitama, Japan
Chapter 6

Steven M. Paul
Neuroscience Discovery Research
Eli Lilly and Co., Lilly Research
    Laboratories
Indianapolis, Indiana, U.S.A.
Chapter 11

Noboru Sasagawa
Department of Life Sciences
Graduate School of Arts and Sciences
University of Tokyo
Meguro-ku, Tokyo, Japan
Chapter 3

Timothy J. Seabrook
Department of Neurology
Center for Neurologic Diseases
Brigham and Women's Hospital
Harvard Medical School
Boston, Massachusetts, U.S.A.
Chapter 12

Edward T. Spooner
Department of Neurology
Center for Neurologic Diseases
Brigham and Women's Hospital
Harvard Medical School
Boston, Massachusetts, U.S.A.
Chapter 12

Beata Szabo
Department of Life Sciences
Graduate School of Arts and Sciences
University of Tokyo
Meguro-ku, Tokyo, Japan
Chapter 3

Sei-ichi Tanuma
Department of Biochemistry
Faculty of Pharmaceutical Sciences
Tokyo University of Science
Shinjuku-ku, Tokyo, Japan
Chapter 3

Michael S. Wolfe
Center for Neurologic Diseases
Brigham and Women's Hospital
Boston, Massachusetts, U.S.A.
email: mwolfe@rics.bwh.harvard.edu
Chapter 4

Berislav V. Zlokovic
Frank P. Smith Laboratory
    for Neurological Surgery
Division of Neurovascular Biology
Center for Aging and Developmental
    Biology
University of Rochester Medical Center
Rochester, New York, U.S.A.
email:
    berislav_zlokovic@urmc.rocheter.edu
Chapter 10

PREFACE

bundant evidence suggests that the buildup and aggregation of amyloid-β (Aβ) peptide in the brain plays a primary role in the pathogenesis of Alzheimer's disease, the major cause of dementia in the elderly. Thus, understanding the metabolism of Aβ is important both for understanding why Aβ accumulates and for development of preventive and therapeutic strategies. This book provides the latest information regarding three major aspects of Aβ metabolism: generation from its precursor, degradation within the brain, and transport out of the brain. All of the authors are internationally known, cutting-edge scientists. This is the first book that specifically, and in detail, covers all aspects of Aβ metabolism. This book will be a resource for graduate students, post-doctoral fellows, and scientists both in this and other disciplines. The contents will hopefully inspire the combination of basic scientists, clinicians, and pharmaceutical leaders to develop new and improved therapies that will have a major impact on a disease that is becoming one of the world's major public health problems.

Takaomi C. Saido

CHAPTER 1

Overview—Aβ Metabolism:

From Alzheimer Research to Brain Aging Control

Takaomi C. Saido

Introduction

Readers and authors of this book alike are the allies of scientists and scientists-to-be in the fight against one of our most common and mightiest enemies, Alzheimer's disease (AD), which deprives individuals of their basic human dignity after decades of (generally) respectful lives with families and friends. As a professional scientist, I am personally grateful to be able to live in the present time when scientists from different ethnic groups, some of which fought against each other in a brutal manner in the past, can now work together to fight the real common enemy through friendly collaborations or in open competition under hopefully true democracies.

However, as authors, we would like readers to know that we do not always share entirely identical ideas or hypotheses until such times as everyone reaches a relevant consensus concerning different aspects of science. We actually need to critically evaluate each other's work for the advancement of science while at the same time maintaining good human relationships (see, for instance, the open debate between Berislav Zlokovic, one of the authors who kindly contributed to this book, and myself).[1] In this respect, not all the chapters here are necessarily consistent with each other, reflecting the differences in opinions. The AD research community is still in the process of proposing different potentially beneficial strategies for the development of preventive and therapeutic measures to combat AD. The ultimate proof of the relevance of any hypothesis or of any experimental results will be real clinical success in a practical manner. Therefore, some of the seemingly relevant strategies are undergoing or will undergo a form of natural selection, such that those, in which the merits outweigh the demerits, will eventually remain. The rest of this book will demonstrate that we are getting closer and closer to clinical success in an accelerated manner, particularly since around the end of the 1980s. The discoveries of gene mutations that cause familial AD (FAD) in the amyloid precursor protein (APP) gene and presenilin genes were the most significant milestones in the 1990s (see section *Etiology of AD* in this chapter and Chapters 2 to 5 for details). (For the nonspecialists, the major primary key abbreviations often used in this book are listed in Table 1.1.)

In any case, I certainly hope that this "scientific globalization," in a positive sense, sharing scientific achievements as cultural common human properties, will be even more improved in the near future because the disease generally does not distinguish between different ethnic groups and because the number of patients in the world will keep growing. Of note is the fact that aging is the strongest risk factor for AD[2] (see section *Towards the Scientific Control of Brain Aging*).

Aβ Metabolism and Alzheimer's Disease, edited by Takaomi C. Saido. ©2003 Eurekah.com.

Table 1.1. Major primary abbreviations used in this book

AD	Alzheimer's disease
FAD*	familial AD
SAD*	sporadic AD
MCI	mild cognitive impairment
FTDP-17	front-temporal dementia with Parkinsonism linked to chromosome 17
NFT	neurofibrillary tangle
ADRD	Alzheimer's disease and Related Disorders (This is the title of the largest international meeting on AD held every other year. The 2002 meeting was held in Stockholm in July 20-25.)
Aβ	amyloid β peptide
APP	amyoid precursor protein (or amyloid protein precursor)
APPs/sAPP	soluble extracellular fragment of APP
APPsα/sAPPα	APPs generated by α-secretase
APPsβ/APPsβ	APPs generated by β-secretase
NTF	N-terminal fragment (of APP generated by α- or β-secretase)
CTF	C-terminal fragment (of APP generated by α- or β-secretase)
C83	CTF generated by α secretase
C99	CTF generated by β secretase
APLP	APP-like protein
NICD	Notch intracellular domain
AICD	APP intracellular domain
ELISA	enzyme-linked immunosorbent assay
CSF	cerebrospinal fluid
ISF	interstitial fluid
BBB	blood-brain barrier
KO	(gene) knock-out
KI	(gene) knock-in
Tg	transgenic
BACE	beta-site APP cleaving enzyme
ADAM	a disintegrin and metalloprotease
PS	presenilin
PKC	protein kinase C
ER	endoplasmic reticulum
TGN	trans-Golgi network
ET	endothelin
IDE	insulin-degrading enzyme
ECE	endothelin-converting enzyme
PDAPP	APP-transgenic mice, in which the transgene expression was driven by platelet-derived growth factor (PDGF) promoter (see ref. 5 in Chapter 11).
ApoE	Apolipoprotein E
CEM	cholesterol-enriched (membrane) microdomain: essentially identical to LDM, DIG, or lipid raft
LDM	(cholesterol- and sphingolipid-rich) low density (membrane) microdomain: essentially identical to CEM, DIG, or lipid raft
DIG	detergent-insoluble glycolipid-rich (membrane domain): : essentially identical to CEM, LDM, or lipid raft
CNS	central nervous system
CAA	cerebral amyloid angiopathy

*FAD is defined as a form of AD that is inherited in an autosomal dominant manner with essentially 100% penetrance. Most FAD cases are early onset, starting between the late 20s and mid 60s. The rest that does not fulfill the definition of FAD is treated as SAD in this book unless otherwise stated. Besides, there has been no genetic risk factor identified that can be used for presymptomatic diagnosis in clinical terms.

For instance, mainland China, with a population of more than one billion people, does not seem to have as high an incidence of AD patients as might be expected from such a large population because the average life span there has been much shorter than in what politicians call more developed countries. However, if China maintains its current rate of industrial and economical growth, this country with its "one-child-per-family" policy, will have not only the largest number but also the highest population ratio of patients with AD and other aging-associated disorders within decades. Unless we do something substantial for the prevention and therapy of these diseases, this scenario would not only result in millions of tragic situations within families but could even induce political destabilization and threaten the peace in some areas, particularly in East Asia, as a result of serious recessions that would be caused by the unexpectedly heavy economic burdens of caring for such people. BUT, if our research efforts can contribute to helping people over the age of 60 stay healthier both physically and intellectually, then society will profit from more active participation of the elderly. This will have benefits for younger generations by reducing the inherent burdens on social welfare systems and on individual care giving, leading to more stable political environments.

The same logic could also apply to other regions of the world. In fact, it is generally established among economists that one major factor which caused recessions in the US and Western Europe in the 1970s to 80s and in Japan in 90s was the change in the ratio of those who needed to be supported economically (and medically) to those who had to support them. (Japan is still struggling with the transition processes even in the 21st century.) Therefore, all of us are responsible not only for the advancement of science but also for the future of human kind. Personally, I wish I could also do something about the other neurodegenerative disorder, Parkinson's disease, too, because my American mother, Dorothy J. Comings, with whom I stayed for one year as a high school exchange student from 1976 to1977, is suffering from this disorder.

Etiology of AD

AD is the major cause of senile dementia in the present world. The estimated number of patients is approximately 20 million worldwide and is expected to keep growing as the world population ages. Now that Mild cognitive impairment[3,4] (MCI), a condition characterized by a significantly reduced memory with cognition being within a normal range, is considered as a prodromal form of dementing disorders primarily represented by AD, the actual number of people being affected by AD pathology is probably much greater than 20 million. Our general understanding is that at the age of 85 one out of every two people is affected either by AD or MCI. In fact, if we consider the temporal distance of decades between the cause and effect in AD pathogenesis, it is possible that some aspects of what has been considered as being part of "normal" brain aging may be prodromal to MCI-associated conditions. (Note that even the autosomal dominantly inherited gene mutations that cause aggressive early onset forms of AD (see the later part of this section) require at least 30-60 years before definitive clinical diagnosis of the onset of the disease can be made.)

Thus, to understand the etiology and mechanism of AD is important not only in conquering this cruel disease but also in realizing the historical dream of human beings to control brain aging (see section *Towards the Scientific Control of Brain Aging*). Figure 1.1 shows a simplified history of AD research since its initial scientific description by Alois Alzheimer in Germany in 1906 (published in 1907).[5] Pathological and patho-biochemical studies mostly up to 1990 established the chronology of the major pathological events at least in neocortex (Fig. 1.2) and identified the molecules comprising the pathological structures. Senile plaques, present extracellularly, consist mainly of amyloid β (Aβ) peptide (Fig. 1.3) while neurofibrillary tangles (NFTs), present intracellularly, contain primarily tau protein. The research community seems to be reaching a consensus that these pathological structures as they appear may not be the direct causes of the symptoms, but rather, that the processes, not fully identified yet, that lead to the pathology may be essential in the pathogenesis.

In any case, as mentioned above, the temporal distance of decades between the cause and effect has been the most challenging factor in AD research. The chronology of pathological events alone does not establish any cause-and-effect relationship. AD research resembles that of archaeology in that researchers need to collect a large body of consistent circumstantial evidence to form a consensus. The observations of pathological structures such as senile plaques and neurofibrillary tangles in AD brain, for instance, have always engendered the arguments as to whether these structures represent pathologically essential and significant pathways or just by-products or consequences of something else essential.

In this respect, identification of the FAD- and tauopathy (FTDP-17: fronto-temporal dementia associated with chromosome 17) causing gene mutations and analyses of their phenotypes in the 1990s have played a major indispensable role in resolving the etiology of AD.[6-8] Consensus has now been reached in that the decades-long cascade leading to dementia is initiated by the deposition of amyloid-β peptide (Aβ) in the brain and that tauopathy is likely to play a major role in the neurodegenerative processes. Thus, the 1990s are called the 'decade of FAD.' I surely hope that the first 10 years of the 2000s will be a decade of SAD, which represents the vast majority of all AD cases.

More than 100 mutations that cause FAD[6,7] have been identified in the three genes encoding the proteins, APP, presenilin 1, and presenilin 2, involved in Aβ generation as shown in Figure 1.4. (See Chapters 2-5 for more details.) A number of papers (probably several hundred) in the 1990s described the phenotype caused by these mutations. Some papers examined the effects of the mutations on Aβ production and others examined effects on cytoskeltal abnormalities involving tau protein, cell death or apoptosis, endoplasmic reticulum (ER)-stresses, etc. These studies have established a consensus that the only phenotype shared by essentially all the mutations in vitro (in cell culture), in vivo (in transgenic and knock-in (KI) mice), and in patients is the increased production of a specific species of Aβ, $Aβ_{1-42}$, which is much more hydrophobic and fibrillogenic than the other major species, $Aβ_{1-40}$. These results have strongly suggested that $Aβ_{1-42}$ is the primary pathogenic agent in the AD cascade. Figure 1.4 shows the processing of APP. Typically, most of the presenilin 1 mutations, accounting for more than 70% of all FAD mutations thus far identified, cause very aggressive presenile Aβ amyoidosis in humans by increasing the steady-state $Aβ_{1-42}$ level approximately 1.5-fold, as detected in the brains of mutant presenilin 1 transgenic or KI mice.[9,10] (See also section on *Towards the Scientific Control of Brain Aging*).

Atypical mutations in the intra-Aβ sequences of APP, such as the Dutch, Flemish, Italian, and Arctic mutations,[6] have also been identified. Most of these mutations result in hemorrhages or strokes caused by unusually severe cerebral Ab amyloid angiopathy (CAA) that also accompanies the presenile parenchymal Aβ deposition. A number of test-tube experiments have demonstrated that these mutations promote Aβ aggregation by altering the peptide conformation.[11-13] Therefore, these rather atypical mutations are generally believed to cause Aβ accumulation through augmenting aggregation[11,12] or protofibril formation.[13] We have recently found that these mutations also cause Aβ to be more resistant to degradation by a physiologically relevant peptidase, neprilysin described in Chapter 6 (Tsubuki S, Takaki Y, Saido TC, submitted for publication). Therefore, these mutations may exert dual pathogenic effects associated not only with aggregation/protofibril formation but also with one major aspect of metabolism, proteolytic degradation.

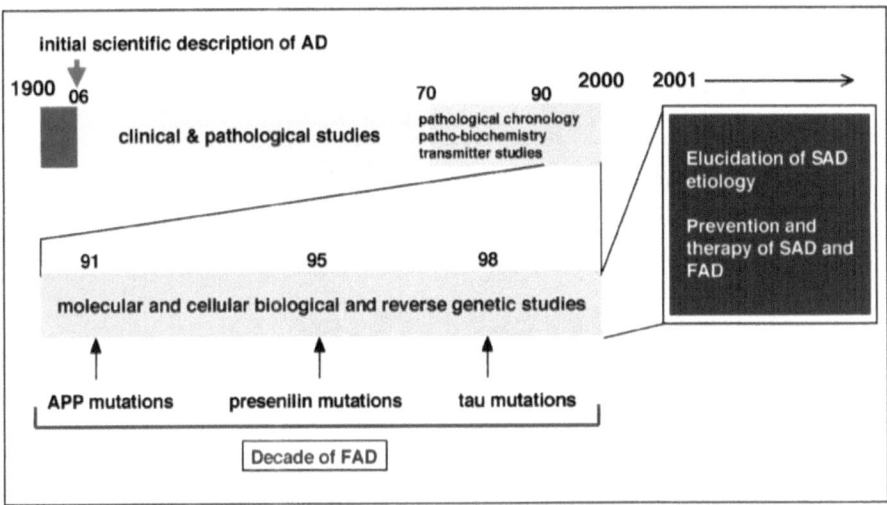

Figure 1.1. Simplified brief history of AD research since 1906.
The major-stream works since the initial clinical description of AD by A. Alzheimer is outlined. The author apologizes for any oversimplification. A hopeful prediction for the future is also included.

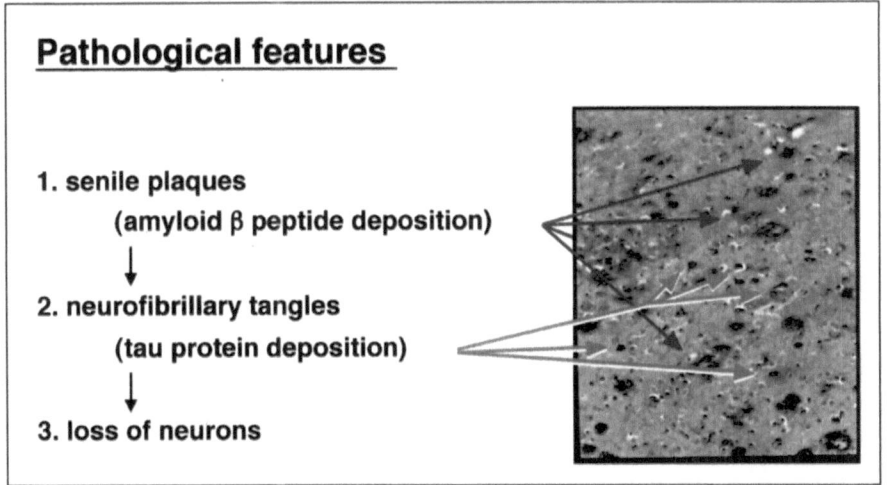

Figure 1.2. Chronology of the major AD pathological events.
The pathological studies mainly in the 1980s established the chronology of the major pathological events at least in neocortex shown in the figure. In fact, the presence of senile plaques, NFTs, and degenerated neurons is necessary for the definitive diagnosis of AD. There has been a consensus that dysfunction or degeneration of neuronal synapses seems to exist between Aβ deposition or PHF formation and neurodegeneration.

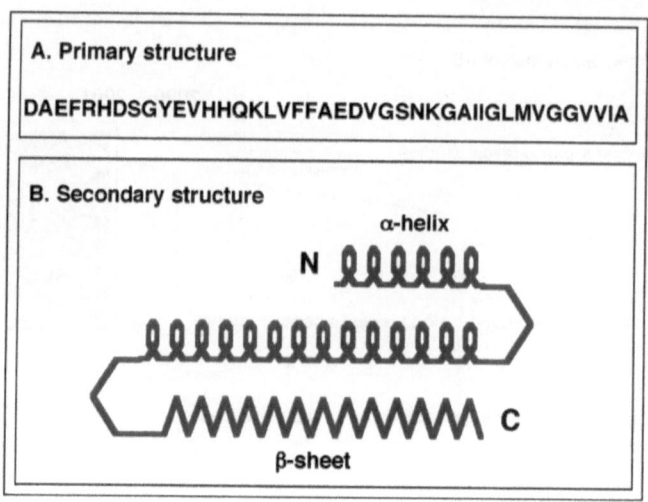

Figure 1.3. Primary and secondary structures of Aβ₁₋₄₂.

The secondary structure was predicted by the Chou-Fasman algorithm using Genentyx software. The molecule is very hydrophobic and has the tendency to aggregate in solution. The α-helix structure in the N-terminal region can easily be converted to β-sheet, making the molecule even more hydrophobic and more apt to aggregate.

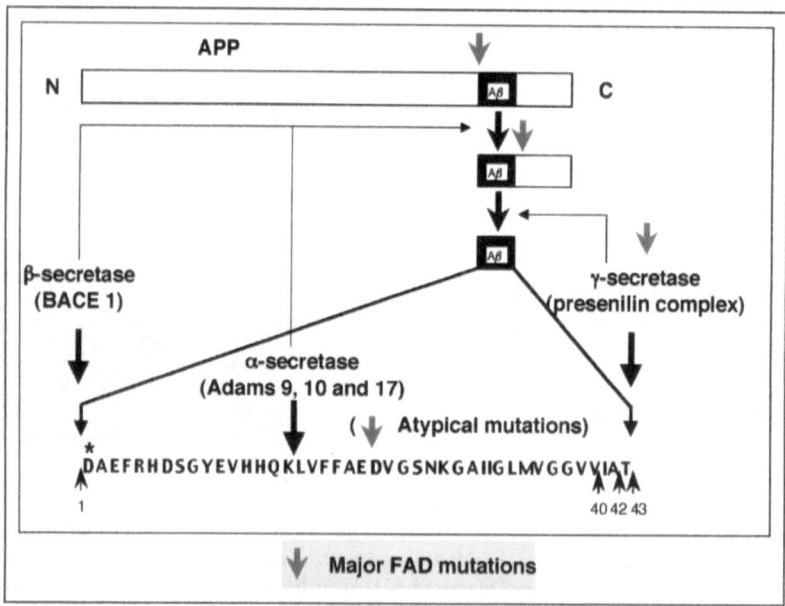

Figure 1.4. Generation of Aβ from its precursor, APP.

APP is first cleaved by β- or α-secretase, generating a C-terminal fragment (C99 or C83), which then is cleaved by γ-secretase to generate Aβ. The major species generated are Aβ1-40 and Aβ1-42. The latter is more hydrophobic and more apt to aggregate and thus is considered to be primarily pathogenic, consistent with the phenotype of the major FAD-causing mutations. The red arrows indicate where in these processes the FAD mutations exert their effects.

Aβ versus Tau

Until some years ago, there used to be arguments between "Baptists" who believed that Aβ was more important and "Tauists" who believed that tau was more important in terms of their contributions to the AD pathogenesis. However, those familiar with the major publications on AD and related disorders in the late 1990s (see the previous section and also Chapters 2-4) do not participate in this kind of discussion any more. I believe that most AD researchers would agree with my view that both are likely to be equally important, particularly in clinical terms. In the AD pathological cascade, it now seems to be just that Aβ is more closely associated with the primary cause while tau is closer to the consequences, such as neurodegeneration. The importance of tau in AD pathogenesis is also apparent from the fact that the quantities of tau accumulated in AD brains are much larger than those in the brains of other neurodegenerative disorders accompanying tauopathy (Taniguchi and Hasegawa, personal communication). Therefore, tau may be a better target for improving the symptoms of patients in a clinically pragmatic manner. Not knowledgeable enough in this specific subject, I will leave this issue to such well-known specialists as J.P. Brion, P. Davies, A. Delacourte, M.. Goedert, M. Hasegawa, Y. Ihara, K Iqbal, V. Lee, Eva & Eckhard Mandelkow, M. Morishima-Kawashima, A. Takashima, R. Terry, J. Trojanowski, C.M. Wischik, the late H.M. Wisniewski (names in alphabetical order), and others (Just net-surf for the reviews and articles under these names), and instead just present my current personal views as follows.

The primary question that is yet to be answered in the domain of tau research is the role of phosphorylation; the tau proteins accumulated in NFT are heavily phosphorylated.[14,15] Although this is an issue beyond the scope of this book on Aβ metabolism, the observations demonstrating that the PHF tau proteins in transgenic mice overexpressing FTDP-17 mutation-carrying tau are also highly phosphorylated[16-18] provide some insights. They seem to logically imply to me that phosphorylation is more likely to be a consequence, possibly of neurons struggling to protect themselves, rather than a cause of NFT formation, unless the mutations exert their effects through altering the phosphorylation/dephosphorylation status of tau, which has never been demonstrated to my knowledge.

Aβ Metabolism: Three Major Targets

Another very important finding in the 1990s is that Aβ is a physiological peptide secreted from neurons under normal conditions both in vitro and in vivo.[19-21] Besides, Aβ does not appear to play a major physiological role; the apparent role of APP processing by the α- and β-secretases is the release of soluble forms of APP, APPs, known for its neuroprotective and neurotrophic functions.[22-24]

Besides, splice variants containing the insert sequences corresponding to the Kunitz-type protease inhibitor (KPI) domain have been identified as protease nexin II, an endogenous inhibitor against a group of serine proteases including thrombin.[25] The other possible function may be the release of a cytoplasmic fragment, generated by γ-secretase, which may translocate to the nucleus and play a regulatory role in transcription in a manner similar to the cleavage of Notch by the γ-secretase activity (See Chapter 5).[26,27] Thus far, several substrates for γ-secretase have been identified. In any of the known activities, the fragment that corresponds to Aβ in APP does not seem to play any major role. Therefore, at present it is most likely that Aβ is simply an unwanted by-product of APP processing.

In any case, APP processing occurs constitutively in the brains of both young and old, and, at least in the brains of young and healthy individuals, no Aβ deposition takes place. Taken all together, these observations clearly indicate that Aβ is constantly anabolized and rapidly catalyzed before being deposited under normal conditions. This catabolism can take place inside the brain or in the circulatory system after transport out of the brain. The kinetic relationships between these three metabolic processes are schematized in Figure 1.5. ($A\beta_{40}$ is left out for the sake of simplicity.)

K_1, K_2, and, K_3 are the rate constants for production, in-parenchyma degradation, and out-of-brain transport of Aβ, respectively. Under the assumptions that the kinetics of the reactions can essentially be analyzed linearly, that these rate constants are independent of each other, and that these processes exist in steady-state equilibrium (see one of the previous reviews for details regarding these assumptions),[28] the relationship between the amounts of $A\beta_{42}$ and APP, represented as $[A\beta_{42}]$ and [APP], respectively, can be expressed by the following equation.

$$[A\beta_{42}] = K_1/(K_2 + K_3) \times [APP] \tag{Formula 1.1}$$

This is based on the following differential equation (See again ref. 31 for details).

$$d[A\beta_{42}]/dt = K_1 \times [APP] - [K_2 + K_3] \times [A\beta_{42}] = 0 \tag{Formula 1.2}$$

A measure of time is expressed as "t" in Formula 1.2. Formula 1.1 is consistent with the phenotypes of almost all the FAD mutations in APP and presenilin 1 genes; K_1 is approximately 1.5-fold greater than that in normal controls, meaning that $[A\beta_{42}]$ also becomes 1.5-fold greater. It also is consistent with one of the phenotypes of Down's syndrome caused by trisomy of chromosome 21 carrying the APP gene; [APP] is 1.5-fold greater than in normal controls and $[A\beta_{42}]$ also becomes 1.5-fold greater.

Therefore, an increase of K_1 (production) or decreases in K_2 (in-parenchyma degradation) and K_3 (out-of-brain transport) can elevate $[A\beta_{42}]$ and thus be causal of pathological Aβ deposition. This logic also indicates that down-regulation of K_1 (production) or up-regulation of K_2 (in-parenchyma-degradation) and K_3 (out-of-brain transport) can decelerate Aβ deposition in the brain and thus will be effective in the prevention and therapy of AD if indeed Aβ plays a primary pathogenic role. Note that the activation of α-secretase(s) would also contribute to reducing K_1 (production) (Chapter 3). The current status of these strategies to achieve the goal of preventing Aβ accumulation is schematized in Figure 1.6. Within the scope of this book, Chapters 1-5 thus focus on the production, 6 and 7 on the in-parenchyma degradation, and 10-12 on the out-of-brain transport. Chapters 9 and 10 describe the possible role of the lipid raft in parenchymal Aβ metabolism and also in pathological Aβ accumulation. Chapter 12 refers to the cellular mechanism of Aβ clearance and to the inflammatory processes associated with Aβ vaccination.

Incidentally, there have been some discussions regarding the relative importance of K_2 (in-parenchyma degradation) versus K_3 (out-of-brain transport) in Aβ clearance.[1] If K_2 is excessively greater than K_3, Formula 1.1 would practically simplify to $[A\beta_{42}] \approx K_1/K_2 \times [APP]$, whereas, if K_2 is excessively smaller than K_3, it would be $[A\beta_{42}] \approx K_1/K_3 \times [APP]$.

I predict a bright future for the control of Aβ levels in the brain through the pursuit of these pathways. Actual approaches and future approaches, based on these strategies, which will have to survive 'natural selection' in a clinical sense (i.e., successes in clinical trials) will be optimally combined so that we will be able to control the Aβ levels in our brains in a manner similar to that of the "cocktail therapy" employed for the treatment of acquired immunodeficiency syndrome (AIDS).[29] In this latter treatment protocol used worldwide, a cocktail of three different anti-human immunodeficiency virus (HIV) strategies suppresses disease development, whereas use of one or two of the three agents generally is ineffective. Moreover, a combination of these anti-Aβ strategies with other strategies such as that targeted at tauopathy will make future prevention and therapy even more promising.

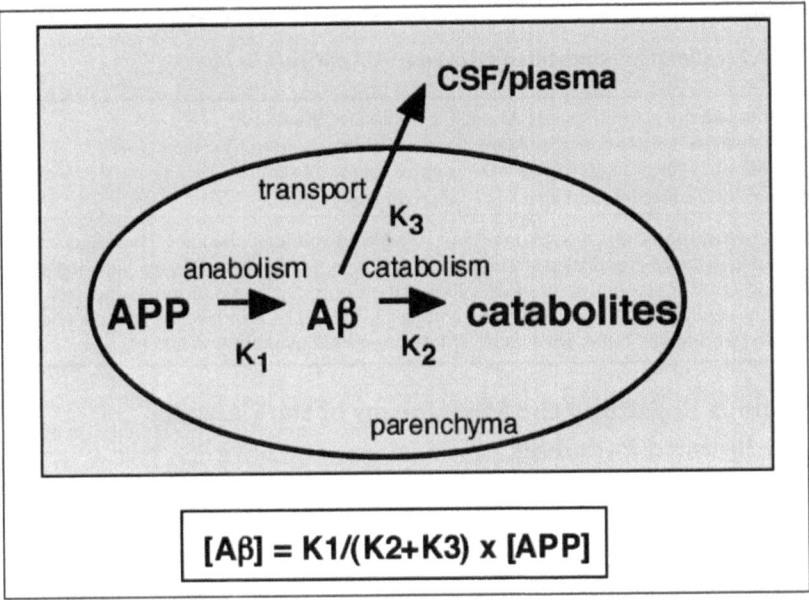

Figure 1.5. Kinetic relationships between production, degradation inside the brain, and transport out of the brain.
The steady-state Aβ (Aβ$_{42}$) level in the brain, [Aβ] ([Aβ$_{42}$]), is primarily a function of the APP level, [APP], the rate constants for production, [K$_1$], in-brain degradation, [K$_2$], and out-of-brain transport, [K$_3$]. See section on Aβ Metabolism: Three Major Targets for details.

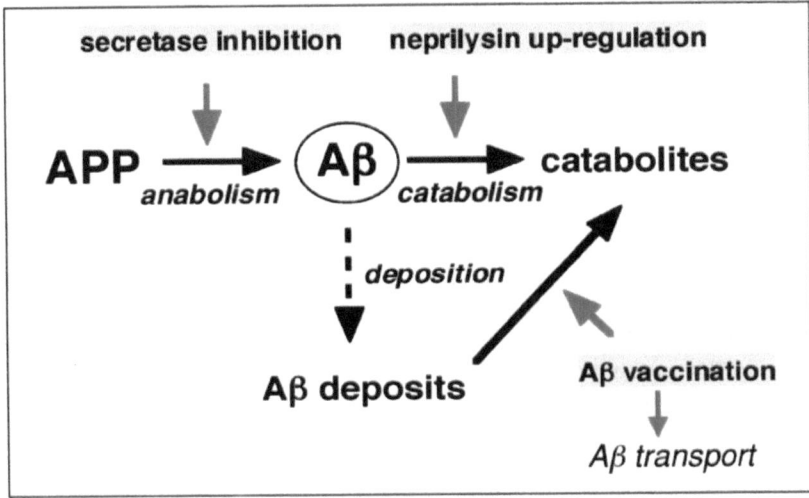

Figure 1.6. A schematized current status of the three major strategies to down-regulate Aβ levels in the brain. Down-regulation of production or up-regulation of in-parenchyma degradation and out-of-brain transport can decelerate Aβ deposition in the brain and thus will be effective in the prevention and therapy of AD if indeed Aβ plays a primary pathogenic role. Note that the activation of α-secretase(s) would also contribute to reducing K$_1$ production (Chapter 3). Aβ vaccination seems to involve both out-of-brain transport (see Chapters 10-12) and cellular disposal of aggregated Aβ (see Chapters 6 and 12 for brief descriptions).

Table 1.2. Estimated numbers of FAD and SAD patients in Japan

Sporadic late-onset (> 60 of age) AD	1-2 million (>99% of all)
Familial late-onset (> 60 of age) AD	not available*
Sporadic early-onset (< 60 of age) AD	approximately 14,000
Familial early-onset (< 60 of age) AD	approximately 1,800

The numbers are based on Campion et al[32] and other epidemiological studies. These figures essentially are doubled in the USA and keep growing as the entire population ages. *To my knowledge, there are no late-onset FAD cases that meet the definition of the autosomal dominant inheritance (see Table 1.1). Having the genotype of apolipoprotein E4 among E2, E3, and E4 is indeed the only genetic risk factor confirmed worldwide, but is not a cause of late-onset FAD by definition herein.

Questions Regarding the Mechanisms of the Cascade of Aβ-Initiated Pathology

One currently unresolved issue is the elucidation of the precise mechanism(s), by which Aβ deposition causes subsequent pathological processes, i.e., tauopathy, dysfunction and degeneration of neurons. This will allow possible opportunities for therapeutic interventions at various time points between Aβ deposition and neurodegeneration. The accelerating effect of excessive amounts of Aβ on tauopathy demonstrated in mouse models[30,31] would indicate the presence of something unknown that relays pathological signals from the former to the latter.

However, perhaps the most important and fundamental question yet to be answered is why Aβ is deposited in SAD which accounts for 99% or more of all AD cases[32] (See Table 1.2.). It should be noted that the number of SAD patients will grow as the average life span increases, whereas the number of FAD patients simply remains proportional to the total population. Because the up-regulation of Aβ production (i.e., increases in [APP] or K_1 in formula 1.1.) is rarely observed prior to the pathological Aβ deposition upon aging, a decrease in the in-parenchyma degradation (K_2) or in the out-of-brain transport (K_3) (or both) is a logical candidate for the primary cause of the majority SAD cases, the expanding burden of aging populations.

Animal Models of AD: The Issues to be Further Addressed

I would like to point out some precautions that we need to bear in mind in relying on research results stemming from the presently available animal models for AD. All the presently available widely accepted Aβ amyloidosis mouse models are transgenic mice that highly overexpress human APP. Although they do reconstitute some of the pathological features of AD (other than Aβ amyloidosis) including synaptic dysfunction and degeneration, dystrophic neurites, inflammatory responses involving activated astrocytes and microglia, and behavioral abnormalities, none of them show the major pathological hallmarks that are actually essential in defining AD: tauopathy and neurodegeneration. The accelerated tauopathy described in the published literature concerning the double APP/tau-transgenic mice[31] is in a sense a matter of course and may simply depend on the high-level expression; even co-overexpression of other proteins such as bovine serum albumin (BSA) may also exert a similar effect if the expressed amount is extremely large. In addition, while the synaptic dysfunction or cognitive deficit has been shown to precede Aβ deposition in the mouse models,[33-35] the AD symptoms become apparent many years after the initial Aβ deposition in humans. Before this deposition, there is little or no apparent sign of synaptic dysfunction in human brains even in those carrying FAD mutations. If it is the "soluble" Aβ oligomers, rather than deposits, that are causing the dysfunction in mice,[36,37] they should be detectable in a well-defined and measurable form that correlates with synaptic dysfunction. To date, this has not been demonstrated to the best of my knowledge. There can be a number of possible reasons accounting for the lack of tauopathy and neurodegeneration in mice as follows.

1. It may be just a matter of time. Mice only live up to three years of age at most, whereas even the most aggressive form of FAD takes at least a couple of decades to present. However, the assumption is not consistent with the fact that APP-transgenic mice accumulate as much Aβ as AD patients do,[37] if Aβ plays the major role in AD pathogenesis.

2. It may be a matter concerning the primary structure of Aβ. In human AD cases, in which the majority (approximately 90%) of both soluble and insoluble Aβ is N-truncated, the most abundant species being $Aβ_{3(pE:\ pyroglutamate)-42}$.[37-39] In contrast, the majority of Aβ accumulated in the mice overexpressing APP is full-length $Aβ_{1-40/42}$.[37] The former species differs from the latter by lacking one positive and two negative charges, thus making it more hydrophobic at the N-terminus than the latter, and thus could be more pathogenic, in a manner analogous to $Aβ_{42}$ versus $Aβ_{40}$.[38,40] Although overexpression of an APP mutant that produces $Aβ_{3(pE:\ pyroglutamate)-42}$ in primary neurons does not seem to have any significant neurodegenerative effect,[41] I believe that the effect must be examined under relevant in vivo conditions.

3. The extent of APP overexpression is unphysiologically high, being several-fold to several tens of times greater than nontransgenic controls. What could one expect if other secretase substrates such as Notch are overexpressed? Most would expect to observe enhanced Notch signaling with an increased release of Notch intracellular domain (NICD); very few would expect any physiological or pathological effect of the overproduced fragment that corresponds to Aβ in APP. There is no reason to reject the same logic from being applied to the APP overexpression paradigm. Indeed, the NICD counterpart fragment of APP, APP-ICD or AICD, has been shown to play a similar transcription-associated role,[26,27] which could be unphysiologically enhanced by APP overexpression. Besides, the primary role of APP processing has been well known for a long time, i.e., release of soluble forms of APP, APP$_s$, which plays neurotrophic and neuroprotective roles[22,23] and also inhibits proteinases such as matrix metalloproteinase (MMP) and a group of serine proteinases including thrombin.[25,42] The presence of too much APP$_s$ may account for the absence of neurodegeneration in these mice.

4. The exaggerated overexpression of APP may also affect the way that APP and other proteins are metabolized. For instance, proteins including Aβ, which are axonally transported to presynapses under normal conditions, may instead undergo endosomal-lysosomal metabolism mainly in the soma. Consistently, the relative increases in the Aβ levels (ratios to the endogenous Aβ) are much smaller than those in APP levels in all of the transgenic mice in my knowledge.

5. Depending on the promoter to drive the expression of transgenic APP, the effects may rather be overly artificial. Neurons are primarily categorized into two major groups, excitatory neurons and inhibitory neurons, and each group is further categorized into a number of subgroups. APP mRNA is not evenly expressed in every neuron and those expressing platelet-derived growth factor (PDGF), prion protein, and thy-1 antigen mRNAs do not probably express the transgene-derived APP in a manner identical to that of the endogenous APP.

6. Humans and mice differ from each other in many ways. A good example is the complement system.[43] Indeed, T. Wyss-Coray demonstrated that inhibition of the complement system led to increased Aβ deposition and occurrence of neurodegeneration.[44] We may thus need to humanize the mice in logically and physiologically relevant manners.

7. Environmental factors represented by diets[45,46] are likely to influence the pathological cascade. The ingredients used in producing experimental mice are different from what humans generally eat.

These and other possible reasons and the lack of NFT and apparent neurodegeneration strongly indicate that the animal models that we have been employing are far from the real model of AD. Therefore, it probably is safe to point out the intrinsic limitation of the use of the present mouse models. Such a careful statement would be even more highly evaluated in the future if, for instance, all vaccination-related data unfortunately turned out to be artifacts that can be observed only in the APP-transgenic mice but not in humans. We need to come up with much more improved models that demonstrate all the major pathological features of AD with minimum artificial manipulation(s). Besides, behind the scenes of the most transgenic studies, one has to make tens, sometimes nearly a hundred, of different lines of mice and choose the ones that show the pathological features seemingly worth publishing. Presence of such processes does not seem to be as scientific as standard science should be because involvement of researchers' wishful thoughts is difficult to fully exclude.

From time to time, the cutting-edge techniques in science keep arising and being refined. Now may be the time when AD researchers fully recognize the consensus of basic cell biologists that overexpression paradigms are often unphysiological and can produce artifacts that may go unnoticed unless there are definitive physiologically relevant controls. Without very careful and logical interpretation and proper controls, the entire research community, led by a relatively small number of influential scientists, could head off in the wrong direction. To perform good science is to discover or to create something new that human beings have never known, not just to put the trendy key words of the era together.

In fact, the term "neurodegeneration" itself is to be outdated for the frontrunners of neuroscience, being too general without specifying which of so many types of neurons undergo neurodegeneration, in what chronology, and through what mechanisms it is preceded by neuronal dysfunction. There already are invisible competitions among frontrunners of neuroscience to be the first to establish each interneuron type-targeted manipulation of specific gene expression for understanding the role of each gene product in both chronological and spatial terms in a relevant in vivo manner.

Moreover, I predict that we will need to introduce the concepts, facts, and methods of Systems Neuroscience to fully understand how mainstream pathological events in the AD cascade lead to the actual symptoms, i.e., loss of memory followed by abnormal reduction of cognition often accompanying psychiatric symptoms such as unusual aggressiveness. I also predict that human genetics to search for genetic risk factors will need to be further developed by establishing more mathematically refined methodology (see Chapter 6.7. for the actual proposal) although I am no specialist in this specific subject.

Towards the Scientific Control of Brain Aging

Recent studies indicate that virtually all humans start to accumulate Aβ in the brain upon aging.[47-50] This suggests that Aβ deposition is an important factor that could determine the life span of our brain and that an AD cascade is initiated by the normal aging process; humans whose "brain life span" is shorter than their "body life span" will suffer from AD in the later stages of their lives. AD has even been described as an "ultimate form of brain aging." This idea provides us with the hope that a large portion of AD cases can be prevented if we can make the brain life span longer than the body life span by reducing Aβ deposition. As outlined in Figure 1.7, most humans start to accumulate Aβ between the ages of 40 and 80.[50-52] This accumulation is initially slow but, after a certain point, deposition begins to accelerate exponentially. Typically, the amounts of $Aβ_{42}$ in the brains of AD patients are 1,000 to 10,000 times higher than in young normal controls. The catastrophic increase implies the presence of multiple positive-feedback vicious cycles originally caused by a slight increase of the steady-state Aβ levels in the brain. Therefore, the profiles of Aβ accumulation in FAD and Down's syndrome

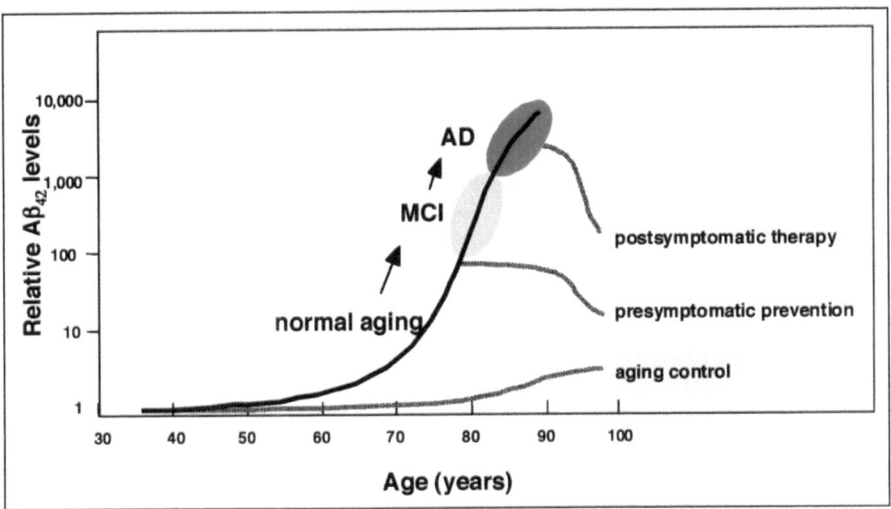

Figure 1.7. Relationship between age, Aβ42 accumulation, normal aging, MCI, and AD.
The Aβ levels in the brain are maintained low in at least young and healthy subjects, but they start to increase at ages between 40 and 80 in most normal people. The increase is generally slow initially but gradually accelerates and eventually reaches a catastrophic situation, where Aβ accumulates in an exponential manner. (Note that the Y-axis is in a log scale.) Typically, the Aβ42 levels in the brains of AD patients are 1,000-10,000-fold higher than in the brains of normal controls.

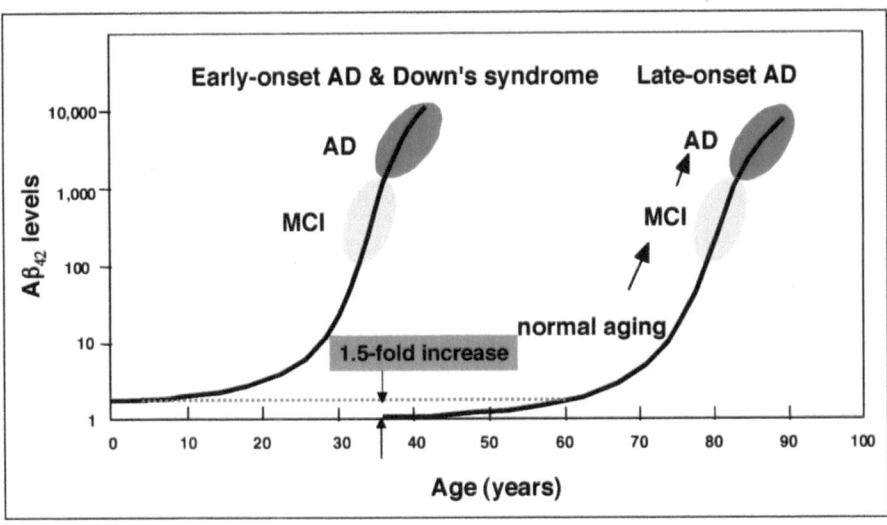

Figure 1.8. FAD (and Down's syndrome) versus SAD in the chronology of Aβ42 accumulation in the brain. The profiles of Aβ accumulation in FAD and Down's syndrome cases being presenile can be explained by considering that the original curve for SAD is shifted to the left as shown in the Figure. The basis of this assumption is an approximately 1.5-fold increase of the steady-state Aβ levels in the brain of FAD and Down's syndrome patients.

cases being presenile can be explained by considering the original curve for SAD is shifted to the left as shown in Figure 1.8.

I am optimistic in a sense that our research efforts will eventually make it possible to control the Aβ levels in our brains. If we can do so in the early stages of AD development, before the massive neurodegeneration takes place, it will serve as a post-symptomatic therapy. (See Fig. 1.7.) If it becomes possible to prediagnose the MCI prodromal to AD, we will be able to initiate a presymptomatic intervention. Because the conversion of what has been interpreted as "normal aging" to AD via MCI appears as a continuous process primarily caused by the gradually accelerating accumulation of Aβ, we may even become able to partially control some very significant aspects of brain aging by maintaining low Aβ levels throughout our lives.

Final Comments

Before closing, I need to inform the readers that I did not succeed in obtaining chapters from some outstanding scientists whose views may differ from some of those included in this book. However, I and other authors have tried to acknowledge their papers and perspectives so that the book covers all the major aspects of Aβ metabolism and adequately represents divergent views of respected researchers. The authors would be more than happy if this book contributes to Alzheimer research by imparting not only the accepted facts, but some of the cutting edge excitement of scientific controversy and inspiring new investigators, particularly young people, to further achievement.

I would like thank all the authors, the very top (relatively young and very active) scientists in the world, and Ronald G. Landes, Landes Bioscience, who led me to edit this book, helped me in many ways. I am very much grateful for Gregory Cole, UCLA, John Trojanowski, University of Pennsylvania, Cynthia Lemere, Harvard Medical School, David Holtzman, Washington University School of Medicine, Maho Morishima-Kawashima, University of Tokyo, and Yasuo Ihara, University of Tokyo, for critical reading of the chapter. I also thank Paul Robbins, University of Pittsburgh, for productive comments on the Introduction (For the acknowledgements for my colleagues, collaborators, etc., see Chapter 6.) I myself have learned a lot through all the editorial procedures, although this book will probably be the last book that I will be editing for many years. Finally, responsibility for all the statements made in this chapter rests with TCS, not with the other authors of this book.

References

1. Zlokovic BV, Yamada S, Holtzman D et al. Clearance of amyloid β-peptide from brain: transport or metabolism? Nat Med 2000; 6(7):718.: Iwata N, Tsubuki S, Hama E et al. Reply to: 'Clearance of amyloid β-peptide from brain: transport or metabolism?' Nat Med 2000; 6(7):718-719.
2. Bick KL, Katzman R. Alzheimer's disease: The road to 2000. Neurosci News 2000; 3(4):6-7.
3. Flicker C, Ferris SH, Reisberg B. Mild cognitive impairment in the elderly: predictors of dementia. Neurology 1991; 41(7):1006-1009.
4. Golomb J, Kluger A, Ferris SH. Mild cognitive impairment: identifying and treating the earliest stages of Alzheimer's disease. Neurosci News 2000; 3(4):46-53.
5. Alzheimer A. Allgem Z Psychiatr-Gerich Med 1907; 64:146-148. Translated version: Schwab C. Neurosci News 2000; 3(4):8-13.
6. Hardy J. APP Mutations Directory: http://www.alzforum.org/members/resources/ app_mutations/index.html
7. Hardy J. Presenilin Mutations Directory: http://www.alzforum.org/members/ resources/ pres_mutations/index.html
8. Kwon J. Tau Mutations Directory: http://www.alzforum.org/res/com/mut/tau/ default.asp
9. Duff K, Eckman C, Zehr C et al. Increased amyloid-β42(43) in brains of mice expressing mutant presenilin 1. Nature 1996; 383(6602):710-713.

10. Nakano Y, Kondoh G, Kudo T et al. Accumulation of murine amyloidβ42 in a gene-dosage-dependent manner in PS1 'knock-in' mice. Eur J Neurosci 1999; 11(7):2577-2581.

11. Murakami K, Irie K, Morimoto A et al. Synthesis, aggregation, neurotoxicity, and secondary structure of various Aβ 1-42 mutants of familial Alzheimer's disease at positions 21-23. Biochem Biophys Res Commun 2002; 294(1):5-10.

12. Demeester N, Mertens C, Caster H et al. Comparison of the aggregation properties, secondary structure and apoptotic effects of wild-type, Flemish and Dutch N-terminally truncated amyloid β peptides. Eur J Neurosci 2001; 13(11):2015-2024.

13. Nilsberth C, Westlind-Danielsson A, Eckman CB et al. The 'Arctic' APP mutation (E693G) causes Alzheimer's disease by enhanced Aβ protofibril formation. Nat Neurosci 2001; 4(9):887-893.

14. Ihara Y, Nukina N, Miura R et al. Phosphorylated tau protein is integrated into paired helical filaments in Alzheimer's disease. J Biochem (Tokyo) 1986; 99(6):1807-1810.

15. Grundke-Iqbal I, Iqbal K, Tung YC et al. Abnormal phosphorylation of the microtubule-associated protein tau (tau) in Alzheimer cytoskeletal pathology. Proc Natl Acad Sci USA 1986; 83(13):4913-4917.

16. Ishihara T, Hong M, Zhang B et al. Age-dependent emergence and progression of a tauopathy in transgenic mice overexpressing the shortest human tau isoform. Neuron 1999; 24(3):751-762.

17. Lewis J, McGowan E, Rockwood J et al. Neurofibrillary tangles, amyotrophy and progressive motor disturbance in mice expressing mutant (P301L) tau protein. Nat Genet 2000; 25(4):402-405.

18. Tanemura K, Murayama M, Akagi T et al. Neurodegeneration with tau accumulation in a transgenic mouse expressing V337M human tau. J Neurosci 2002; 22(1):133-141.

19. Shoji M, Golde TE, Ghiso J et al. Production of the Alzheimer amyloid β protein by normal proteolytic processing. Science 1992; 258:126-129.

20. Seubert P, Vigo-Pelfrey C, Esch F et al. Isolation and quantification of soluble Alzheimer's β-peptide from biological fluids. Nature 1992; 359:325-327.

21. van Gool WA, Schenk DB, Bolhuis PA. Concentrations of amyloid-β protein in cerebrospinal fluid increase with age in patients free from neurodegenerative disease. Neurosci Lett 1994; 172(1-2):122-124.

22. Saitoh T, Sundsmo M, Roch JM et al. Secreted form of amyloid β protein precursor is involved in the growth regulation of fibroblasts. Cell 1989; 58(4):615-622.

23. Mattson MP, Cheng B, Culwell AR et al. Evidence for excitoprotective and intraneuronal calcium-regulating roles for secreted forms of the β-amyloid precursor protein. Neuron 1993; 10(2):243-254.

24. Kang J, Lemaire HG, Unterbeck A et al. The precursor of Alzheimer's disease amyloid A4 protein resembles a cell-surface receptor. Nature 1987; 325(6106):733-736.

25. Oltersdorf T, Fritz LC, Schenk DB et al. The secreted form of the Alzheimer's amyloid precursor protein with the Kunitz domain is protease nexin-II. Nature 1989; 341(6238):144-147.

26. Kimberly WT, Zheng JB, Guenette SY et al. The intracellular domain of the β-amyloid precursor protein is stabilized by Fe65 and translocates to the nucleus in a notch-like manner. J Biol Chem 2001; 276(43):40288-40292.

27. Leissring MA, Murphy MP, Mead TR et al. A physiologic signaling role for the γ-secretase-derived intracellular fragment of APP. Proc Natl Acad Sci USA 2002; 99(7):4697-4702.

28. Saido TC. Alzheimer's disease as proteolytic disorders: anabolism and catabolism of β-amyloid. Neurobiol Aging 1998; 19(1 Suppl):S69-75.

29. Flexner C. Dual protease inhibitor therapy in HIV-infected patients: pharmacologic rationale and clinical benefits. Annu Rev Pharmacol Toxicol 2000; 40:649-674.

30. Götz J, Chen F, van Dorpe J et al. Formation of neurofibrillary tangles in P301L tau transgenic mice induced by Aβ42 fibrils. Science 2001; 293:1491-1495.

31. Lewis J, Dickson DW, Lin WL et al. Enhanced neurofibrillary degeneration in transgenic mice expressing mutant tau and APP. Science 2001; 293:1487-1491.

32. Campion D, Dumanchin C, Hannequin D et al. Early-onset autosomal dominant Alzheimer disease: prevalence, genetic heterogeneity, and mutation spectrum. Am J Hum Genet 1999; 65(3):664-670.

33. Hsia AY, Masliah E, McConlogue L, et al. Plaque-independent disruption of neural circuits in Alzheimer's disease mouse models. Proc Natl Acad Sci USA 1999; 96(6):3228-3233.

34. Moechars D, Dewachter I, Lorent K et al. Early Phenotypic Changes in Transgenic Mice That Overexpress Different Mutants of Amyloid Precursor Protein in Brain. J Biol Chem 1999; 274:6483-6492.

35. Mucke L, Masliah E, Yu GQ et al. High-level neuronal expression of Aβ1-42 in wild-type human amyloid protein precursor transgenic mice: synaptotoxicity without plaque formation. J Neurosci 2000; 20:4050-4058.

36. Walsh DM, Klyubin I, Fadeeva JV et al. Naturally secreted oligomers of amyloid β protein potently inhibit hippocampal long-term potentiation in vivo. Nature 2002; 416(6880):535-539.

37. Kawarabayashi T, Younkin LH, Saido TC et al. Age-dependent changes in brain, cerebrospinal fluid, and plasma amyloid β protein in the Tg2576 transgenic mouse model of Alzheimer's disease. J Neurosci 2001; 21:372-381.

38. Saido TC, Iwatsubo T, Mann DMA et al. Dominant and differential deposition of distinct β-amyloid peptide species, AβN3(pE), in senile plaques. Neuron 1995; 14:457-466.

39. Russo C, Schettini G Saido TC et al. Preferential deposition of truncated amyloid-β peptides in brain of presenilin 1 gene mutation carriers. Nature 2000;, 405: 531-532.

40. Jarrett JT, Lansbury Jr PT. Seeding "one-dimensional crystallization" of amyloid: a pathogenic mechanism in Alzheimer's disease and scrapie? Cell 1993; 73(6):1055-8.

41. Shirotani K, Tsubuki S, Lee HJ et al. Generation of amyloid β peptide with pyroglutamate at position 3 in primary cortical neurons. Neurosci Lett 2002; 327(1):25-28.

42. Miyazaki K, Hasegawa M, Funahashi K et al. A metalloproteinase inhibitor domain in Alzheimer amyloid protein precursor. Nature 1993; 362(6423):839-841.

43. Akiyama H, Barger S, Barnum S et al. Inflammation and Alzheimer's disease. Neurobiol Aging 2000; 21(3):383-421.

44. Wyss-Coray T, Yan F, Lin AH et al. Prominent neurodegeneration and increased plaque formation in complement-inhibited Alzheimer's mice. Proc Natl Acad Sci USA 2002; 99(16):10837-10842.

45. Behl C, Davis J, Cole GM et al. Vitamin E protects nerve cells from amyloid β protein toxicity. Biochem Biophys Res Commun 1992; 186(2):944-950.

46. Lim GP, Chu T, Yang F et al. The curry spice curcumin reduces oxidative damage and amyloid pathology in an Alzheimer transgenic mouse. J Neurosci 2001; 21(21):8370-8377.

47. Funato H, Yoshimura M, Kusui K et al. Quantitation of amyloid β-protein (Aβ) in the cortex during aging and in Alzheimer's disease. Am J Pathol 1998; 152:1633-1640.

48. Funato H, Enya M, Yoshimura M et al. Presence of sodium dodecyl sulfate-stable amyloid β-protein dimers in the hippocampus CA1 not exhibiting neurofibrillary tangle formation. Am J Pathol 1999; 155(1):23-28.

49. Morishima-Kawashima M, Oshima N, Ogata H et al. Effect of apolipoprotein E allele ε4 on the initial phase of amyloid β-protein accumulation in the human brain. Am J Pathol 2000; 157:2093-2099.

50. Wang J, Dickson DW, Trojanowski JQ, Lee VM. The levels of soluble versus insoluble brain Abeta distinguish Alzheimer's disease from normal and pathologic aging. Exp Neurol 1999; 158:328-337.

β-Secretase: Progress and Open Questions

Martin Citron

Abstract

Finding inhibitors of Aβ42 generation is a major goal of Alzheimer's disease drug development. Two target protease activities, β-and γ-secretases, were operationally defined more than 10 years ago, but progress in this area has been slow because the actual enzymes were not identified. Using an expression cloning strategy we have identified a novel membrane bound aspartic protease, BACE1, as β-secretase. This finding has been confirmed and BACE1 and its homologue BACE2 have been characterized in detail by many groups. Major progress has been made in two areas: First, the x-ray crystal structure, which is critical for rational inhibitor design, has been solved and shown to be similar to that of other pepsin family members. Second, knockout studies show that BACE1 is critical for Aβ generation, but the knockout mice show an otherwise normal phenotype, raising the possibility that therapeutic BACE1 inhibition could be accomplished without major mechanism-based toxicity. However, target-mediated toxicity of β-secretase inhibition cannot be ruled out based on the currently available data alone. While various peptidic β-secretase inhibitors have been published, the key challenge now is the generation of more drug-like compounds that could be developed for therapeutic purposes. Other current areas of investigation, including identification of BACE1 substrates and the potential role of BACE1 overexpression in AD, are discussed.

Identification of β-Secretase

Since the cloning of amlyloid precursor protein (APP) revealed that Aβ must be excised from the middle of its large precursor protein,[1] the two necessary proteolytic cleavage events, one at the N-terminus by an enzyme termed "β-secretase" and one at the C-terminus by an enzyme termed "γ-secretase", have attracted a lot of attention. This is understandable because Aβ formation is the initial step in the hypothetical amyloid cascade[2] and is thus supposed to be ultimately responsible for the pathology of Alzheimer's disease (AD). Moreover, the necessity of proteolytic cleavage for Aβ generation immediately suggested the existence of two potential therapeutic intervention targets which could be addressed using standard protease inhibition approaches. Consequently, APP processing and Aβ generation have been studied in a variety of systems by many investigators and their results are summarized in Figure 2.1. At least three distinct protease activities are involved in processing the membrane protein APP along two major pathways, the α-secretase and the amyloid forming β-secretase pathway. A relatively small minority of APP molecules enters the β-secretase pathway in which β-secretase cleaves APP and releases a soluble fragment, β-APPs. The C-terminal membrane bound C99 peptide is then cleaved by γ-secretase within the transmembrane domain and two major isoforms of 40 and 42 amino acid length with different C-termini, Aβ$_{40}$ and Aβ$_{42}$, are generated. In the α-secretase pathway, α-secretase cleaves in the middle of the Aβ region (thus precluding Aβ

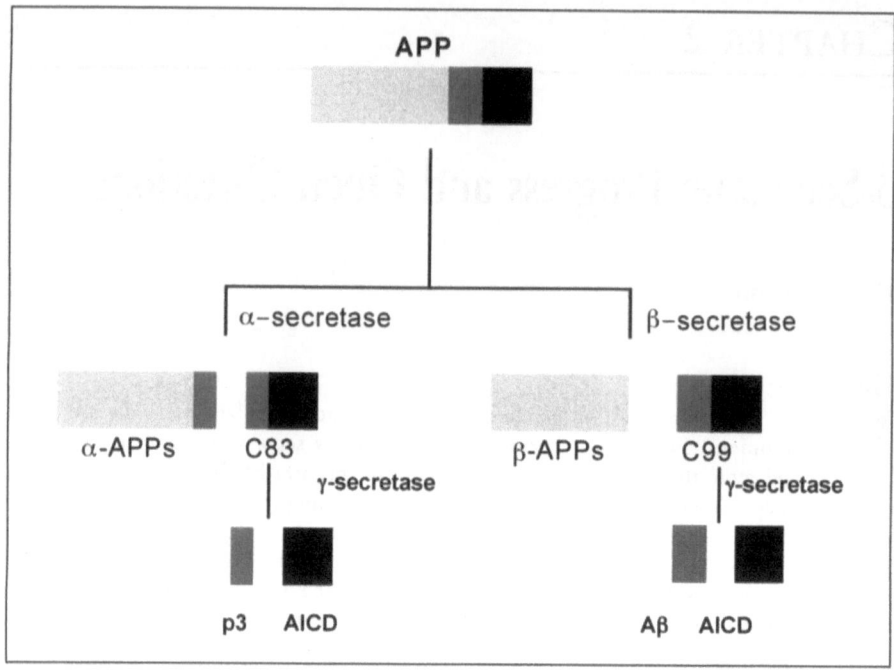

Figure 2.1. Overview of APP processing. APP is shown with the large N-terminal ectodomain in light grey, the Aβ region in dark grey, and the C-terminal amyloid intracellular domain (AICD) in black (not drawn to scale). APP can be processed by α-secretase to yield secreted α-APPs and the membrane bound C83 fragment or by β-secretase to yield β-APPs and the membrane bound C99 fragment. These membrane bound fragments can each undergo γ-secretase cleavage to give rise to the secreted fragments Aβ and p3 and AICD which may play a role in signaling (discussed in other chapters in this book).

formation) and releases a soluble fragment, α-APPs. The remaining membrane-bound C-terminal fragment C83 is then cleaved by γ-secretase to give rise to p3$_{40}$ and p3$_{42}$, shortened versions of Aβ that do not appear to be major plaque components. Pharmacological and cell biological studies demonstrated early on that the three major activities, α- β- and γ-secretase were distinct. By the mid 1990s an approximate subcellular localization of these activities was established by defining where one can detect their respective cleavage products. Using this approach it was for example shown that β-secretase activity must reside both in endosomes[3] and in the secretory pathway,[4] whereas α-secretase activity could be clearly detected on the cell-surface.[5] However, it was far less clear whether all cleavages within each major class were carried out by one or several different, but related enzymes. Obviously, such questions could be addressed directly, and inhibitors could be developed in a rational way, if the relevant enzymes were identified and isolated. Therefore, isolating β-and γ-secretases has been a major goal for laboratories in academia and the pharmaceutical industry for a long time. It took more than 10 years to accomplish this, primarily because the biochemical purification of these enzymes using peptidic substrates proved exceedingly difficult. Upon homogenization of cells numerous active enzymes capable of cleaving peptidic substrates at the right position are released and a variety of them have been suggested as candidates for β- and/or γ-secretase over the years (for a detailed review of these efforts from the mid-90s see ref. 6). Clearly, there was a problem with irrelevant enzymes performing artifactual cleavages of short peptidic substrates that obscured the less robustly expressed secretases. We decided to circumvent the intrinsic problems of a biochemical purification approach by using an expression cloning strategy to identify genes that modulate Aβ production. We hypothesized that overexpression of α secretase in cells overexpressing APP could lead to increased Aβ production. A cDNA library was prepared from 293 human embryonic kidney cells (which are known to express the complete APP processing machinery) and divided into pools of 100, and these pools were transfected into 293 human embryonic kidney cells overexpressing APP. Pools causing increased Aβ production were then subdivided into smaller pools and finally reduced to single clones. This expression cloning strategy ultimately led to the identification of beta-site APP-cleaving enzyme 1 (BACE1) as the major β-secretase.[7] Subsequently three other groups reported isolation of the same enzyme using different approaches. Hypothesizing that β-secretase belongs to the aspartic protease family and using a genomics approach and antisense studies, Yan et al isolated β-secretase.[8] In contrast, Sinha et al used biochemical affinity purification to identify the enzyme.[9] Finally, Hussain et al also reported identification of β-secretase, but they did not report why the particular candidate was selected initially.[10]

Characterization of β-Secretase

BACE1 is a 501 amino acid protein with an aminoterminal signal peptide of 21 amino acids followed by a proprotein domain spanning amino acids 22 to 45 (Fig. 2.2). The lumenal domain of the mature protein extends from residues 46 to 460 and is followed by a transmembrane domain of 17 residues and a short cytosolic tail of 24 amino acids. BACE1 contains two active site motifs at amino acids 93 to 96 and 289 to 292 in the lumenal domain, each containing the highly conserved signature sequence of aspartic proteases D T/S G T/S. Based on the amino acid sequence, BACE1 is predicted to be a type I transmembrane protein with the active site on the lumenal side of the membrane where β-secretase cleaves APP. At the amino acid level, BACE1 shows less than 30% sequence identity with human pepsin family members. The BACE1 gene is localized to chromosome 11q23.3, not associated with AD.

We have thoroughly demonstrated that BACE1 exhibits all the known properties of β-secretase[7] and subsequent publications from other groups have confirmed this analysis. Tissue culture and animal studies indicated that β-secretase is expressed in all tissues, but higher in neurons of the brain. This is exactly what we found for BACE1 mRNA. However, we and others also found very high levels of BACE1 mRNA in the pancreas, leading to speculations about a role of BACE1 as a protease in this tissue, but recently this mRNA has been shown to encode a shortened BACE1 splice variant of unknown function which is deficient in protease activity.[11] We also demonstrated the presence of BACE1 protein in brain.[7] BACE1 has the right topological orientation to attack the β-secretase cleavage site of APP and is localized within acidic intracellular compartments, such as endosomes and trans-Golgi network (TGN) as expected for β-secretase.[7] BACE1 is also detectable at the cell surface and cycling of the protein between the cell membrane and endosomes was documented.[12,13] Endosomal targeting of BACE1 depends on a cytoplasmic dileucine motif (residues 499 and 500).[12] Endogenous BACE1 was shown to primarily localize to the TGN from which a small portion is delivered to the plasma membrane from which it then recycles to endocytic compartments. It appears that the BACE1 transmembrane domain contains a TGN targeting signal.[14] Overexpression of BACE1 in cells increases the β-secretase products C99 (the major β-cleavage product starting with the Asp1 amino acid of Aβ), C89 (the β'-cleavage product starting with the Glu11 amino acid of Aβ) and APPsβ, while the α-secretase product APPsα is decreased. In cells expressing wild-type APP this directly leads to increased Aβ generation.[7] Under overexpression conditions the ratio of β'/β-cleavage correlates with BACE1 expression levels,[15] both APP and C99 can undergo β'-cleavage[16] and β-site proteolysis predominates in the endoplasmic reticulum, whereas β'-cleavage predominates in the trans-Golgi network.[17] Antisense inhibition of BACE1 decreases β-secretase cleavage and Aβ generation.[7,8] Purified forms of

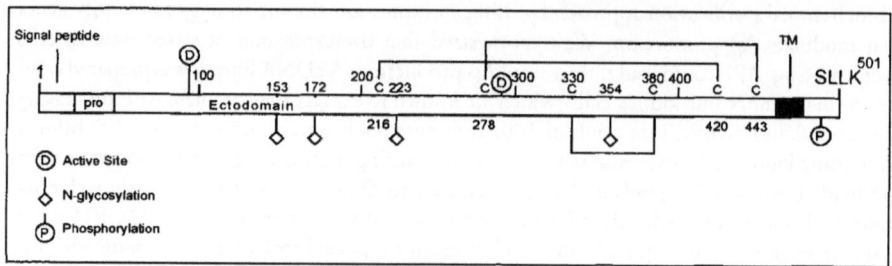

Figure 2.2. Schematic representation of BACE1, the β-secretase protein (drawn to scale). BACE1 is initially synthesized as a 501 amino acid transmembrane protein. The signal peptide and the propeptide are indicated. Active site aspartate residues are marked. The transmembrane domain is indicated in black. The four N-glycosylation sites are marked and the connectivity of the intramolecular disulfide bonds of the extracellular domain is shown. The last 4 amino acids are shown. Serine 498 is the cytoplasmic phosphorylation site. The dileucine motif is an endosomal targeting signal.

BACE1 cleave APP substrates in vitro at the correct site and with the same P1 specificity previously described for β-secretase in cell-based assays. Purified forms of BACE1 have an acidic pH optimum and are not inhibited by pepstatin, as expected for β-secretase. BACE1 undergoes a series of posttranslational modifications. All four potential N-glycsosylation sites in the ectodomain are occupied by carbohydrate; no O-glycosylation is detected.[18] Protease activity is affected by the occupancy of glycosylation sites, but the effect is likely to be indirect, e.g., by enhancing correct folding or increasing solubility of the protein.[19] The six cysteine residues in the ectodomain all form intramolecular disulfide bonds in a pattern which is not conserved in other aspartic proteases.[18] In pulse chase experiments BACE1 protein (calculated MW 50 kDa) is initially detectable as a 60 kDa immature glycosylated form made in the ER which undergoes rapid maturation to a 70 kDa product which is stable.[13,18] Ectodomain shedding of BACE1 upon overexpression has been reported.[20] It is not known whether this ectodomain shedding occurs under physiological conditions and its functional significance remains unclear. BACE1 is initially synthesized as a proprotein that is cleaved at residue E46 to form the mature enzyme. Interestingly, both the proprotein and the mature protein are proteolytically active and the Pro domain does not suppress activity as in a strict zymogen, but appears to facilitate proper folding of the active enzyme.[21] The prodomain cleavage is not autocatalytic, as in some other aspartic proteases. Instead, furin, or a furin-like protease is likely responsible for the propeptide cleavage.[22] After full maturation BACE1 can be phosphorylated within its cytoplasmic domain at Ser498. The phosphorylation regulates retrieval of BACE1 from endocytosed vesicles, a mechanism reminiscent of furin trafficking.[23]

BACE2, the Closest Relative

Immediately after the identification of BACE1 database mining led to the discovery of BACE2, an aspartic protease which has 64% amino acid sequence similarity to BACE1 and which also shows a C-terminal transmembrane domain. Together BACE1 and BACE2 define a novel family of aspartic proteases (Fig. 2.3). BACE2 localizes to the obligate Down's syndrome region of chromosome 21, making it attractive to speculate that it may have β-secretase activity and that its overexpression could contribute to the AD pathology observed in Down's syndrome. However, its brain expression is very low and it is primarily peripherally expressed.[24] Several other studies also indicate that this enzyme does not play a major role in Aβ generation. First, in contrast to BACE1, antisense inhibition does not impact Aβ generation.[8] Second, overexpression of BACE2 does not lead to increased Aβ production[24-26] and very recent data

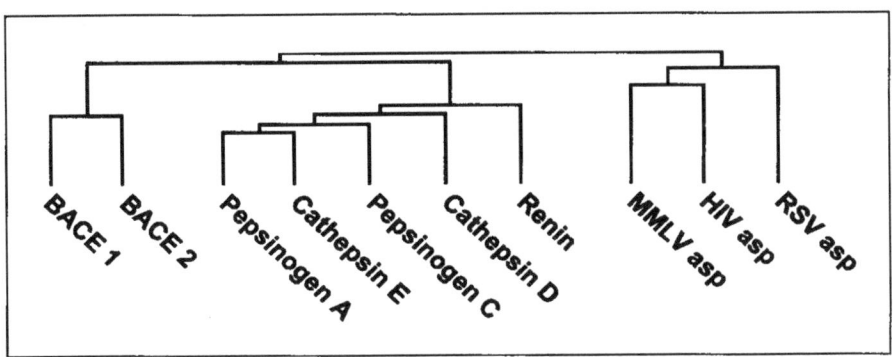

Figure 2.3. Evolutionary tree showing the relationships between BACE1, BACE2 and other aspartic proteases. BACE1 and BACE2 form a novel family of transmembrane aspartic proteases that are most closely related to the pepsin family. The homodimeric retroviral aspartic proteases are evolutionarily more ancient than the BACE and pepsin families.

suggest that BACE2 may even have an antiamyloidogenic role.[27] While BACE2 does recognize the β-site of APP, it primarily cleaves APP after the known α-secretase sites Phe-19 and Phe-20 which does not lead to increased Aβ generation.[26,28] Third, BACE1 knockout alone is sufficient to abolish Aβ production (see below). Currently the physiological role of BACE2 and its substrates are unknown, but because it is closely related to the important BACE1, it is interesting to analyze the properties of BACE2. Like BACE1, BACE2 undergoes posttranslational modifications and is transported through the secretory pathway to the plasma membrane, but in contrast to BACE1 it is hard to detect in endosomes.[29] The transmembrane domain of BACE2 is only 22% identical to that of BACE1 and this may explain why BACE2 is not primarily localized to the TGN, but has a more diffuse localization pattern.[28] In contrast to BACE1, BACE2 prodomain processing is autocatalytic.[28,30] The crystal structure of BACE2 has not been disclosed and the phenotype of BACE2 knockout mice has not been reported yet.

BACE1 Transgenics and Knockouts

Mice transgenic for human BACE1 expressed under the mouse Thy-1 promoter have been generated. High BACE1 protein expression in the hippocampus and cortex was observed and stable lines were established, suggesting that neuronal BACE1 overexpression in vivo is not lethal. Detailed pathology of the transgenics has not been reported, and the focus of the study was on showing that in vivo BACE1 overexpression leads to increased amyloidogenic processing of APP, as demonstrated by reduced levels of full-length APP and increased levels of C99, C89 and APPsβ, ultimately resulting in increased Aβ production.[31] This result does not come as a surprise, but was expected based on the tissue culture data. A recent report from another group suggests that BACE1 overexpression under the same promoter leads to a neurodegeneration phenotype with motor deficits.[32] It is not yet known why this phenotype was not observed in the first study. For inhibitor development, knockout mouse studies are more important, and consequently knockout mouse results were available earlier than transgenic data. The finding that β-secretase knockout mice are deficient in Aβ production, independently reported by us and two other groups, was not unexpected, but it did provide ultimate in vivo validation of BACE1 as β-secretase and it demonstrated that in mice no compensatory mechanism for β-secretase cleavage exists.[33-35] The more exciting and unexpected aspect of the knockout studies was the absence of major problems due to β-secretase ablation: We analyzed the phenotype of our BACE1 knockout mice and found them to be healthy and fertile. A detailed analysis demonstrated that the knockout mice are normal in terms of gross morphology and anatomy, tissue histology, hematology and clinical chemistry.[33] In addition, behavioral analysis of the knockout mice generated by Roberds et al showed no obvious deficits in basal neurological and physiological functions.[35] While no study has investigated aged knockout mice or knockout mice subjected to various challenges, the absence of distinct pathology is very encouraging for β-secretase drug development.

β-Secretase Drug Development Is Under Way

On theoretical grounds inhibition of either β- or γ-secretase should be sufficient to block Aβ production (see Fig. 2.1), so the choice between these two targets could then be determined by technical feasibility and mechanism-based toxicity issues. In the absence of purified secretase enzymes, whole cell assays for inhibition of Aβ generation in the presence of compounds were run, and these assays delivered potent γ-secretase inhibitor leads, but no potent β-secretase inhibitors were disclosed in the scientific or patent literature. Consequently, during the 90s most drug development focused on the γ-secretase target. The recent insights into the potential liabilities of γ-secretase inhibition and the availability of pure β-secretase enzyme have led to a surge in interest in β-secretase inhibitor development. BACE1 is an attractive drug target: Its biology is relatively well understood, the knockout data show that the enzyme is necessary and sufficient for β-secretase cleavage and, most importantly, they suggest that mechanism based toxicity of inhibitors may be nonexistent or manageable, and other aspartic proteases have been successfully targeted for drug development in the past.[36] Moreover, BACE1 is predominantly expressed in the brain and because only very few aspartic proteases exist, the risk for cross-inhibition is limited. The crystal structure of the BACE1 ectodomain complexed with a peptidic inhibitor has been solved and this structural information is now available to guide inhibitor development. The overall structure of the enzyme is very similar to that of other known aspartic proteases. However, there are differences in the details of the active site which is generally more open and less hydrophobic than in other aspartic proteases.[37] Potent peptidic inhibitors of BACE1 have already been described in several publications.[9,38-40] The goal of ongoing studies is the design of smaller, more drug-like compounds. A recent kinetic study using synthetic peptide libraries and mass spectrometry for initial rate determination has led to a clearer understanding of the subsite specificity of BACE1. It was shown that a consensus peptide defined by the kinetic study as EIDLMVLD was cleaved with a kcat/KM value 14-fold better than the analogous APPsw derived octamer. These data can be used to design better substrates and inhibitors. They also suggest that much better physiological substrates for β-secretase than APP may exist.[41] The relatively large active site of β-secretase may pose challenges for the development of small molecule inhibitors. Based on the experience with other aspartic protease inhibitors, in particular renin and human immunodeficiency virus (HIV) protease drugs, one can predict that the generation of drug-like β-secretase inhibitors will be difficult and time consuming, but that it will be pursued by many companies, given the size of the AD problem and the limited number of validated alternative AD targets.

Controversies and Open Questions

Compared to other areas of AD research, the β-secretase field has seen surprisingly little controversy. I would define three major landmarks of the field for which all publications agree or—if there is only one publication—everyone seems to agree with the published data and their interpretation. These landmarks are identification of BACE1 as the major β-secretase activity, the discovery that BACE1 knockout mice do not make Aβ, but seem otherwise normal and the description of the crystal structure of the enzyme. Several publications describing the detailed properties of BACE1 are also in agreement with each other with some minor differences on subcellular localization. While there is no more doubt about the identity of the long-sought after β-secretase, there are still numerous questions left to answer, and some of these questions may impact β-secretase inhibitor development: First, what are the other substrates of BACE1? Additional substrates are likely to exist, because it is counterintuitive to assume that β-secretase has evolved just to generate Aβ. The above mentioned kinetic study suggesting that physiological β-secretase substrates could have several-hundred-fold better k_{cat}/K_m values than APP clearly supports this notion.[41] This is interesting from a basic perspective, but also important to know in order to predict potential liabilities of β-secretase inhibitors. So far, two new potential β-secretase substrates, ST6Gal I, a sialyltransferase[42] and most recently p-selectin glycoprotein ligand-1,[43] have been proposed, mainly based on the finding that their secretion was increased upon BACE1 overexpression. It remains to be determined whether these proteins are actually cleaved by BACE1 at physiological expression levels. Second, what is the biological role of BACE2? What is the phenotype of a BACE2 knockout? What is the phenotype of a BACE1,2 double knockout? Again this is interesting from a basic perspective, but it would also be important to know, whether a β-secretase inhibitor drug would have to be BACE1 specific or whether cross-inhibition of BACE2 will be acceptable. Third, does overexpression of BACE1 play a role in some forms of AD? Three recent reports suggest that this may be the case,[44-46] but more work is needed to find out whether BACE1 overexpression can be firmly linked to AD. Fourth, how is β-secretase expression regulated? This area still appears largely unexplored. Fifth, can β-secretase activity be modulated by mechanisms other than direct enzyme inhibition? A provocative recent paper claims that BACE1 partitions into lipid rafts and that this may underlie the cholesterol sensitivity of Aβ production, thus potentially linking BACE1 with the cholesterol field.[47] And finally, the ultimate question addressing the amyloid hypothesis remains to be asked in the clinic—assuming a safe β-secretase inhibitor drug can be found, will it prevent AD, will it arrest progression or will it even give symptomatic improvement?

References

1. Kang J, Lemaire HG, Unterbeck A et al. The precursor of Alzheimer's disease amyloid A 4 protein resembles a cell-surface receptor. Nature 1987; 325:733-736.
2. Hardy J, Allsop D. Amyloid deposition as the central event in the aetiology of Alzheimer's disease. Trends in Pharmac 1991; 12:383-388.
3. Koo EH, Squazzo S. Evidence that production and release of amyloid β-protein involves the endocytic pathway. J Biol Chem 1994; 269:17386-17389.
4. Haass C, Lemere CA, Capell A et al. The Swedish mutation causes early-onset Alzheimer's disease by β-secretase cleavage within the secretory pathway. Nature Med 1995; 1:1291-1296.
5. Sisodia SS. β-amyloid precursor protein cleavage by a membrane-bound protease. Proc Natl Acad Sci USA 1992; 89:6075-6079.
6. Evin G, Beyreuther K, Masters CL. Alzheimer's disease amyloid precursor protein (AβPP): proteolytic processing, secretases and βA4 amyloid production. Amyloid: Int J Exp Clin Invest 1994; 1:263-280.
7. Vassar R, Bennett BD, Babu-Khan S et al. β-secretase cleavage of Alzheimer's amyloid precursor protein by the transmembrane aspartic protease BACE. Science 1999; 286:735-741.
8. Yan R, Bienkowski MJ, Shuck ME et al. Membrane-anchored aspartyl protease with Alzheimer's disease β-secretase specificity. Nature 1999; 402:533-537.
9. Sinha S, Anderson JP, Barbour R et al. Purification and cloning of amyloid precursor protein β-secretase from human brain. Nature 1999; 402:537-540.
10. Hussain I, Powell D, Howlett DR et al. Identification of a novel aspartic protease (Asp2) as β-secretase. Mol Cell Neurosci 1999; 14:419-427.
11. Bodendorf U, Fischer F, Bodian D et al. A splice variant of β-secretase deficient in the amyloidogenic processing of the amyloid precursor protein. J Biol Chem 2001; 276:12019-12023.

12. Huse JT, Pijak DS, Leslie GJ et al. Maturation and endosomal targeting of β-site amyloid precursor protein-cleaving enzyme. J Biol Chem 2000; 275:33729-33737.

13. Capell A, Steiner S, Willem M et al. Maturation and pro-peptide cleavage of β-secretase. J Biol Chem 2000; 275:30849-30854.

14. Yan R, Han P, Miao H et al. The transmembrane domain of the Alzheimer's β-secretase (BACE1) determines its late Golgi localization and access to β-amyloid precursor protein (APP) substrate. J Biol Chem 2001; 276:36788-36796.

15. Creemers JW, Dominguez DI, Plets E et al. Processing of β-secretase by furin and other members of the proprotein convertase family. J Biol Chem 2001; 276:4211-4217.

16. Liu K, Doms RW, Lee VMY. Glu11 site cleavage and N-terminally truncated Aβ production upon BACE overexpression. Biochemistry 2002; 41:3128-3136.

17. Huse JT, Liu K, Pijak DS et al. β-secretase processing in the trans-Golgi network preferentially generates truncated amyloid species that accumulate in Alzheimer's disease brain. J Biol Chem 2002; 277:16278-16284.

18. Haniu M, Denis P, Young Y et al. Characterization of Alzheimer's β-secretase protein BACE A pepsin family member with unusual properties. J Biol Chem 2000; 275:21099-21106.

19. Charlwood J, Dingwall C, Matico R et al. Characterization of the glycosylation profiles of Alzheimer's β-secretase protein Asp-2 expressed in a variety of cell lines. J Biol Chem 2001; 276:16739-16748.

20. Benjanne S, Elagoz A, Wickham L et al. Post-translational processing of β-secretase (β-amyloid-converting enzyme) and its ectodomain shedding. J Biol Chem 2001; 276:10879-10887.

21. Shi XP, Chen E, Yin KC et al. The pro domain of β-secretase does not confer strict zymogen-like properties but does assist proper folding of the protease domain. J Biol Chem 2001; 276:10366-10373.

22. Bennett BD, Denis P, Haniu M et al. A furin-like convertase mediates propeptide cleavage of BACE, the Alzheimer's β-secretase. J Biol Chem 2000; 275:37712-37717.

23. Walter J, Fluhrer R, Hartung B et al. Phosphorylation regulates intracellular trafficking of β-secretase. J Biol Chem 2001; 276:14634-14641.

24. Bennett BD, Babu-Khan S, Loeloff R et al. Expression analysis of BACE2 in brain and peripheral tissues. J Biol Chem 2000; 275:20647-20651.

25. Hussain I, Powell DJ, Howlett DR et al. Asp1 (BACE2) cleaves the amyloid precursor protein at the β-secretase site. Mol Cell Neurosci 2000; 16:609-619.

26. Farzan M, Schnitzler CE, Vasilieva N et al. BACE2, a β-secretase homolog, cleaves at the β-site and within the amyloid-β region of the amyloid-β precursor protein. Proc Natl Acad Sci 2000; 97:9712-9717.

27. Basi G, Frigon N, Tatsuno G, Xu M et al. Cellular knock-out of endogenous BACE2 reveals its role in APP processing homeostasis of Aβ production in cells coexpressing BACE1 and BACE2. Neurobiol Aging 2002; 23:S178-S179.

28. Yan R, Munzner JB, Shuck ME et al. BACE2 functions as an alternative α-secretase in cells. J Biol Chem 2001; 276:34019-34027.

29. Fluhrer R, Capell A, Westmeyer G et al. A non-amyloidogenic function of BACE-2 in the secretory pathway. 2002.

30. Hussain I, Christie G, Schneider K et al. Prodomain processing of Asp1 (BACE2) is autocatalytic. J Biol Chem 2001; 276:23322-23328.

31. Bodendorf U, Danner S, Fischer F et al. Expression of human β-secretase in the mouse brain increases the steady-state level of β-amyloid. J Neurochem 2002; 80:799-806.

32. Rockenstein E, Mante M, Adame A et al. Overexpression of human beta-secretase (BACE1) results in severe motor deficits and neurodegeneration in tg mice. Neurobiol Aging 2002; 23:S235-S236.

33. Luo Y, Bolon B, Kahn S et al. Mice deficient in BACE1, the Alzheimer's β-secretase, have normal phenotype and abolished β-amyloid generation. Nat Neurosci 2001; 4:231-232.

34. Cai H, Wang Y, McCarthy D et al. BACE1 is the major β-secretase for generation of Aβ peptides by neurons. Nat Neurosci 2001; 4:233-234.

35. Roberds SL, Anderson J, Basi G et al. BACE knockout mice are healthy despite lacking the primary β-secretase activity in brain: implications for Alzheimer's disease therapeutics. Hum Mol Genet 2001; 10:1317-1324.

36. Leung D, Abbenante G, Fairlie DP. Protease inhibitors: current status and future prospects. J Med Chem 2000; 43:305-341.
37. Hong L, Koelsch G, Lin X et al. Structure of the protease domain of memapsin 2 (β-secretase) complexed with inhibitor. Science 2000; 290:150-153.
38. Ghosh AK, Shin D, Downs D et al. Design of potent inhibitors for human brain memapsin 2 (β-secretase). J Am Chem Soc 2000; 122:3522-3523.
39. Ghosh AK, Bilcer G, Harwood C et al. Structure-based design: potent inhibitors of human brain memapsin 2 (β-secretase). J Med Chem 2001; 44:2865-2868.
40. Tung JS, Davis DL, Anderson JP et al. Design of substrate-based inhibitors of human β-secretase. J Med Chem 2002; 45:259-262.
41. Turner RT, Koelsch G, Hong L et al. Subsite specificity of memapsin 2 (β-secretase): implications for inhibitor design. Biochemistry 2001; 40:10001-10006.
42. Kitazume S, Tachida Y, Oka R et al. Alzheimer's β-secretase, β-site amyloid precursor protein cleaving enzyme, is responsible for cleavage secretion of a Golgi-resident sialyltransferase. Proc Natl Acad Sci 2001; 98:13554-13559.
43. Lichtenthaler S, Seed B. The aspartyl protease BACE regulates the shedding of p-selectin glycoprotein ligand-1. Neurobiol Aging 2002; 23:S175-S176.
44. Holsinger RMD, McLean CA, Masters CL et al. BACE and β-secretase product CTFβ are increased in sporadic Alzheimer's disease brain. Neurobiol Aging 2002; 23:S177.
45. Shen Y, Yang L-B, Yang X-L et al. Alteration of beta secretase (BACE) expression in sporadic Alzheimer's brains. Neurobiol Aging 2002; 23:S190.
46. Fukamoto H, Cheung B, Hyman BT et al. β-site amyloid precursor protein cleaving enzyme (BACE) activity is increased in temporal neocortex of Alzheimer's disease. Neurobiol Aging 2002; 23:S181.
47. Riddell DR, Christie G, Hussain I et al. Compartmentalization of β-secretase (Asp2) into low-buoyant density non-caveolar lipid rafts. Curr Biol 2001; 11:1288-1293.

CHAPTER 3

APP α-Secretase, a Novel Target for Alzheimer Drug Therapy

Shoichi Ishiura, Masashi Asai, Chinatsu Hattori, Nika Hotoda, Beata Szabo, Noboru Sasagawa and Sei-ichi Tanuma

Abstract

The neurodegeneration in Alzheimer's disease (AD) may be caused by deposition of amyloid β peptide (Aβ) in plaques in brain tissue (amyloid hypothesis). Mechanisms of Aβ production in the brain have been the subject of considerable interest. Several factors regulate processing of amyloid precursor protein (APP), which is cleaved by three types of membrane-bound proteases designated α- β- and γ-secretases. Although we are only beginning to understand their functions, the discovery of nonamyloidogenic α–secretase has already provided some answers to longstanding questions of Alzheimer drug development. Here we summarize recent advances in the identification of APP α-secretases. Pharmacological up-regulation γ-secretases should provide rational drug design for Alzheimer's disease.

Introduction

Dementia can be produced by a large number of pathological processes. Alzheimer's disease is the most common form of senile dementia and is characterized by the progressive impairment of cognitive domains caused by a loss of neurons from particular regions of the cerebral cortex, accompanied by the presence of amyloid deposition. A major component of senile plaques is amyloid β-protein (Aβ) and this 39-43 amino acid peptide plays a crucial role in the pathogenesis of Alzheimer's disease. Aβ is derived by proteolysis of the amyloid precursor protein (APP)[1-4] a ubiquitously expressed type I transmembrane protein. But in normal brain, Aβ deposition can not be found even in aged people. APP undergoes endoproteolytic processing within the Aβ domain in the trans-Golgi network (TGN) and at the cell surface (Fig. 3.1), and the generation of Aβ is inhibited. The α-secretase cuts the extracellular domain closer to the membrane, leaving 12 amino acids on the external surface. This results in the "constitutive" secretion of the large extracellular domain of APP (sAPP) into the medium. It has been shown that the fraction of sAPP can be increased by activating protein kinase C (PKC) cascade ("regulated" α-cleavage). Stimulation of α-cleavage of APP leads to a significant decrease in Aβ formation. Three related metalloproteases seem to exert an α-secretase activity. Therefore, stimulation of α-secretase as well as inhibition of amyloidogenic enzymes (β- and γ-secretases) have been suggested to theoretically reduce Aβ production.[5,6]

Aβ Metabolism and Alzheimer's Disease, edited by Takaomi C. Saido. ©2003 Eurekah.com.

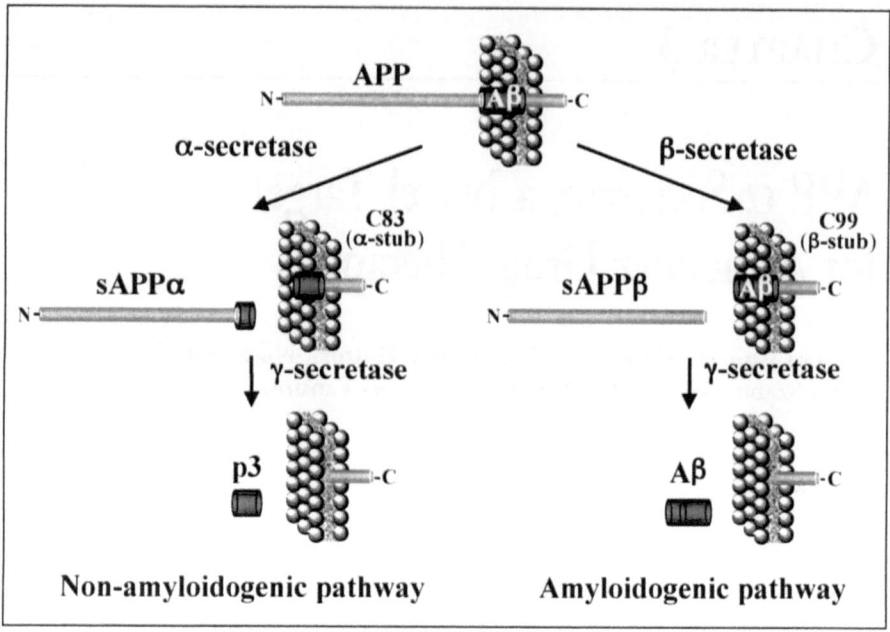

Figure 3.1. Proteolysis of APP.

ADAMs and α-Secretase

Several lines of evidence suggest that α-secretase activity is modulated by metal ions and metalloprotease inhibitors. Three members of ADAM (a distintegrin and metalloprotease) family (Table 3.1), ADAM9 (meltrin γ MDC9, ADAM10 (MADM) and ADAM17 (TNFα-converting enzyme or TACE), are reported to be candidate α-secretases (Table 3.2). These have the consensus sequence of zinc-dependent metalloprotease "HEXGHXXGXXHD" and seem to be active metalloprotease. In 1998, TACE was reported to play a central role in regulated α-cleavage by the use of gene disruption.[7] Primary embryonic fibroblasts derived from TACE-knockout mice lost regulated sAPP secretion, while basal or constitutive secretion was not affected. Recent results of Slack et al demonstrate that constitutive sAPP secretion was also increased in HEK293 cells transiently transfected with TACE cDNA.[8] These indicate that TACE is capable of catalyzing constitutive α-secretory cleavage of APP, but additional members of the metalloprotease, possibly ADAMs, mediate endogenous constitutive cleavage of APP.

In 1999, Lammich et al reported that transfection of ADAM10 cDNA into HEK293 cells increased basal and PKC-stimulated α-secretase activity.[9] Several investigators confirm the results.[10-13] We independently demonstrated in 1999 that ADAM9 has an α-secretase activity in COS cells.[14] ADAM9 and 10 were localized by immunostaining and cell surface biotinylation in the plasma membrane and Golgi apparatus. These results support the view that APP is cleaved both at the cell surface and along the secretory pathway. Hooper et al suggested that constitutive α-cleavage of APP occurred in a post-TGN,[6] while it is primarily an intracellular source that is cleaved in the regulated pathway. One possible explanation for the uncertainty about the identity of α-secretase is that there may be more than one enzyme depending on the cell type and on the cellular localization.

ADAM9 (MDC9) has been suggested to participate in the shedding of the ectodomain of the heparin-binding epidermal growth factor (HB-EGF) in response to phorbol esters. We

Table 3.1. ADAM family

No	alternative nomenclature(s)		No	alternative nomenclature(s)	
1	fertilin α, PH-30 α	+	18	tMDC III, ADAM27	-
2	fertilin β, PH-30 β	-	19	meltrin β	+
3	Cyritestin, tMDC I, CYRN1	-	20		+
4	tMDC V	-	21	ADAM31	+
5	tMDC II	-	22	MDC2 α / β	-
6	tMDC IV	-	23	MDC 3	-
7	EAP-1, GP-83	-	24	Testase 1	+
8	MS2, CD156	+	25	Testase 2	+
9	meltrin γ, MDC9, MCMP, KIAA0021	+	26	Testase 3	+
10	MADM, kuzbanian (kuz), sup-17	+	27	tMDC III, ADAM18	-
11	MDC	-	28	eMDC II, MDC-Lm, MDC-Ls	+
12	meltrin α, MCMP, MLTN, MLTNA	+	29	svph1	-
13	xMDC13	+	30	svph4	+
14	adm-1	-	31	ADAM21	+
15	metargidin, MDC15, AD56, CRII-7	+	32	AJ131563	-
16	MDC16	+	33	ADAM13	+
17	TACE, cSVP	+	34	Testase 4	+

+: proteolytic activity confirmed; -: not confirmed.

found that coexpression of mouse ADAM9 (mADAM9) and APP cDNA into COS cells enhanced both constitutive and regulated α-cleavage of APP.[15] When we transfected APP695-LAA in which HHQK of Aβ13-16 is replaced with LHHAA to confer greater resistance to α-secretory cleavage, phorbol ester failed to activate α-cleavage, indicating that regulated cleavage is conducted by ADAM9. We also found that a general metalloprotease inhibitor SI-27 inhibited the phorbol ester-induced increase in sAPPα secretion together with concomitant increase in sAPPβ secretion. Increasing concentration of the inhibitor had reciprocal effects on sAPPα and sAPPβ secretion.[14] Purified recombinant mADAM9 could cleave APP in vitro at α-cleavage site.

To clone human counterpart of mADAM9, total RNA was prepared from human glioblastoma cell line A172 which shows a high endogenous α-secretase activity. cDNA was synthesized with the Thermoscript RT-PCR System (Life Technologies). One clone had a divergent carboxyl terminus and was designated hADAM9s.[16] A full-length human ADAM9 (hADAM9) consists of an open reading frame of 2460nt encoding 819 amino acid residues. The former lacks 106bp, resulting in premature termination (total 655 amino acids) and the deletion of the C-terminal transmembrane domain. We isolated this truncated hADAM9s protein from the culture medium of cDNA-transfected COS cells and showed that hADAM9s was capable of processing APP at α-site. All these results strongly suggest that α-secretory cleavage occurs at the cell surface and modulation of APP metabolism can be achieved from extracellular space.

Some members of the ADAM family (ADAMs 11, 12, 28) have alternative splicing forms and have been secreted. ADAM12s is expressed in some tumor cell lines and placenta, while full-length ADAM12 is expressed in every tissue. ADAM12s is found in pregnancy serum and cleaves the major IGF-binding protein in human serum, IGFBP-3. These results suggest that the splicing form of ADAM has tissue-specific unique functions.

Recently mice lacking ADAM9 were generated.[17] They developed normally, are viable and fertile, and did not have any pathological phenotypes. Interestingly, there were no differences in the production of the Aβ and p3 (APP α- γ-secretase product as well as HB-EGF

Table 3.2. Comparison Among ADAMs 9, 10 and 17

	ADAM9	ADAM10	ADAM17
aliases	MDC9, meltrin γ	MADM, kuzbanian (kuz)	TACE
chromosome	8p11.21	15q22	2p25
organism besides Homo sapiens [percent identity and length of aligned region]	Bos taurus Caenorhabditis elegans [37%,641aa] Drosophila melanogaster [37%,207aa] Mus musculus [85%,818aa] Rattus norvegicus [31%,610aa] Sus scrofa Xenopus laevis	Bos taurus Caenorhabditis elegans [43%,715aa] Drosophila melanogaster [46%,584aa] Mus musculus [26%,531aa] Rattus norvegicus [97%,543aa] Sus scrofa Xenopus laevis	Drosophila melanogaster [31%,471aa] Mus musculus [25%,539aa] Rattus norvegicus [35%,421aa] Sus scrofa
length of mRNA	3,865bp	3,410bp	3,014bp
length of Protein	819aa	748aa	824aa
sequence of catalytic active site	HELGHNLGMNHD	HEVGHNFGSPHD	HELGHNFGAEHD
physiological substrates besides APP	insulin B-chain HB-EGF pro TNF-α	type IV collagen myelin basic protein CD40 pro TNF-α	p55 TNFR I p75 TNFR II CD30 L-selectin interleukin-6 receptor pro TGF-α pro TNF-α

shedding. This argues against an essential role for ADAM9 as an α-secretase in mice (see Future Perspectives).

Relations between α- and Other Secretases

Aβ peptide is generated through the proteolytic cleavage of the APP molecule by two proteases, termed β- and γ-secretases. β-secretase the N-terminus of Aβ to produce a soluble form of APP (sAPPβ) and a 99-residue C-terminal membrane-bound fragment C99. C99 is substrate for membrane-bound γ-secretase, which clips in the middle of the transmembrane region. Candidate β-secretase was identified as BACE1, a membrane-bound aspartic protease.

Recent results from Selkoe laboratory suggest that presenilins 1 and 2, which are endoplasmic reticulum (ER) or Golgi protein and whose dysfuction is a cause of early-onset autosomal dominant cases of familial Alzheimer's disease, are atypical aspartic proteases involved in the processing of APP and Notch.[18] Since presenilins do not seem to fit into any known family of protease, some investigators cast doubts on the proteolytic nature of presenilins[19-21] and they argue that presenilins are simply important components of a high-molecular-weight γ-secretase complex and act like a cofactor or a molecular chaperone of γ–secretase. Purification of γ-secretase complex to homogeneity could solve this problem.

Future Perspectives

To identify the endogenous α-secretase in A172 cells, RNA interference (RNAi) was induced by double-stranded RNA (dsRNA) encoding ADAM sequences. Recently it was been reported that specific gene-silencing with 21-nucleotide dsRNAs in mammalian cells provides a useful and reasonable tool. A172 cells are exceptional human glioblastoma cells with an endogenous potent α-secretase activity. A single application of RNAi decreased the amount of sAPPα in the medium by 25% compared with the control, and double RNAi and triple RNAi caused a 63-77% and 87% suppression of α-secretase activity, respectively (Asai M et al, manuscript in preparation). The results indicate that ADAM9, ADAM10 and ADAM17 catalyze α-secretory cleavage and therefore act as α-secretases in A172 cells.

Despite our lack of full knowledge concerning the identity of APP α-secretase, it is clear that the activation of these proteases may offer new therapeutic methods. Inhibitors of β- and γ-secretases may prevent the deposition of Aβ in brain, whereas an activator of α-secretase (and Aβ-degrading enzymes) may inhibit the accumulation of Aβ with the same efficiency.

There is strong evidence to suggest that NSAIDs may prevent or delay the onset of Alzheimer's disease. Indomethacin at 25-50 mM promoted nonamyloidogenic α-secretase in A172 cells.[22] Therefore NSAIDs may provide an avenue to prevention of the Alzheimer's disease.

Acknowledgments

This work was supported in part by a Grant-in-Aid for Scientific Research on Priority Areas(C)-Advanced Brain Science Project-from the Ministry of Education, Culture, Sports, Science and Technology, Japan.

References

1. Hardy J, Selkoe DJ. The amyloid hypothesis of Alzheimer's disease: Progress and problems on the road to therapeutics. Science 2002; 297:353-356.
2. Mills J, Reiner PB. Regulation of amyloid precursor protein cleavage. J Neurochem 1999; 72:443-460.
3. Brown MB, Ye J, Rawson RB et al. Regulated intramembrane proteolysis: A control mechanism conserved from bacteria and humans. Cell 2000; 100:391-398.
4. Nunan J, Small DH. Regulation of APP cleavage by α-, β-, and γ-secretases. FEBS Lett 2000; 483:6-10
5. Ishiura S, Hotoda N, Szabo B et al. Alzheimer's amyloid α-secretase: A special target for drug development. Psychogeriatrics 2001; 1:273-276.
6. Hooper NM, Turner AJ. The search for α-secretase and its potential as a therapeutic approach to Alzheimer's disease. Curr Med Chem 2002; 9:1107-1119.
7. Buxbaum JD, Liu KN, Luo Y et al. Evidence that tumor necrosis factor α converting enzyme is involved in regulated α-secretase cleavage of the Alzheimer amyloid protein precursor. J Biol Chem 1998; 273:27765-27767.

8. Slack BE, Ma LK, Seah CC. Constitutive shedding of the amyloid precursor protein ectodomain is up-regulated by tumour necrosis factor-α converting enzyme. Biochem J 2001; 357:787-794.

9. Lammich S, Kojro E, Postina R et al. Constitutive and regulated α-secretase cleavage of Alzheimer's amyloid precursor protein by a disintegrin metalloprotease. Proc Natl Acad Sci USA 1999; 96:3922-3927.

10. Lopez-Perez E, Zhang Y, Frank SJ et al. Constitutive α-secretase cleavage of the β-amyloid precursor protein in the furin-deficient LoVo cell line: involvement of the pro-hormone convertase 7 and the disintegrin metalloprotease ADAM10. J Neurochem 2001; 76:1532-1539.

11. Anders A, Gilbert S, Garten W et al. Regulation of the α-secretase ADAM10 by its prodomain and proprotein convertases. FASEB J 2001; 15:1837-1839.

12. Marcinkiewicz M, Seidah NG. Coordinated expression of β-amyloid precursor protein and the putative β-secretase BACE and α-secretase ADAM10 in mouse and human brain. J Neurochem 2000; 75:2133-2143.

13. Kojro E, Gimpl G, Lammich S et al. Low cholesterol stimulates the nonamyloidogenic pathway by its effect on the α-secretase ADAM 10. Med Sci 2001; 98:5815-5820.

14. Koike H, Tomioka S, Sorimachi H et al. Membrane-anchored metalloprotease MDC9 has an α-secretase activity responsible for processing the amyloid precursor protein. Biochem J 1999; 343:371-375.

15. Hotoda N, Koike H, Sasagawa N et al. Amyloid precursor protein is secreted by membrane-anchored metalloprotease ADAM9. In: Mizutani S, ed. Cell Surface Aminopeptidases. Elsevier Science B.V. 2001:401-406.

16. Hotoda N, Koike H, Sasagawa N et al. A secreted form of human ADAM9 has an α-secretase activity for APP. Biochem Biophys Res Commun 2002; 293:800-805.

17. Weskamp G, Cai H, Brodie TA et al. Mice lacking the metalloprotease-disintegrin MDC9 (ADAM9) have no evident major abnormalities during development or adult life. Mol Cell Biol 2002; 22:1537-1544.

18. Wolfe MS, Xia W, Ostaszewski BL et al. Two transmembrane aspartates in presenilin-1 required for presenilin proteolysis and γ–secretase activity. Nature 1999; 398:513-517.

19. Steiner H, Haass C. Intramembrane proteolysis by presenilins. Nature Rev Mol Cell Biol 2000; 1:217-224.

20. Small DH. The role of presenilins in γ-secretase activity: catalyst or cofactor? J Neurochem 2001; 76:1612-1614.

21. Checler F. The multiple paradoxes of presenilins. J Neurochem 2001; 76:1621-1627.

22. Kinouchi T, Ono Y, Sorimachi H et al. Arachidonate metabolites affect the secretion of an N-terminal fragment of Alzheimer' disease amyloid precursor protein. Biochem Biophys Res Commun 1995; 209:841-849.

Chapter 4

γ-Secretase and Presenilin

Michael S. Wolfe

Abstract

Because production and deposition of the amyloid-β peptide (Aβ) is intimately linked to the pathogenesis of Alzheimer's disease (AD), the proteases responsible for excising Aβ from the amyloid-β precursor protein (APP), β- and γ-secretases, are considered important therapeutic targets. β-secretase is a membrane-anchored aspartyl protease in the pepsin family. In contrast, γ-secretase is a highly complex and unusual protease that catalyzes hydrolysis within the transmembrane domain of its substrates. A large body of evidence now supports the hypothesis that the catalytic component of γ-secretase is presenilin, a multi-pass membrane protein. Genetic analysis identified presenilins as major loci for mutations that cause familial, early onset AD. These mutations affect Aβ production by altering the specificity of γ-secretase, and knockout of presenilins eliminates γ-secretase activity. The identification of transition-state analogue inhibitors of γ-secretase suggested an aspartyl protease mechanism. Consistent with all these findings, two conserved transmembrane aspartates in presenilin are critical for γ-secretase activity, and active site directed inhibitors of γ-secretase bind directly to presenilin. Moreover, presenilin copurifies with γ-secretase activity through size exclusion and affinity chromatography, and antibodies to presenilin can precipitate γ-secretase activity. Presenilins by themselves do not possess protease activity. Instead, presenilins are apparently part of a larger protease complex that includes the single-pass protein nicastrin and at least two other components.

Introduction

Ever since the discovery that certain dominant missense mutations in the amyloid-β precursor protein (APP) cause early-onset Alzheimer's disease (AD)[1,2] and alter production of the amyloid-β peptide (Aβ),[3-6] the two proteases responsible for excision of Aβ from APP have been considered prime targets for therapeutic intervention for the prevention and/or treatment of AD. These proteases, β- and γ-secretases, have been the subject of intense study toward identification and characterization in the hope that such understanding would facilitate drug discovery efforts. β-secretase prunes APP on its lumenal/extracellular side near the transmembrane region, leading to shedding of the ectodomain and the formation of the N-terminus of Aβ on the membrane-bound C-terminal remnant. β-secretase was identified in 1999 as a novel membrane-tethered aspartyl protease in the pepsin family.[7-10] A soluble form of β-secretase has been cocrystallized with an inhibitor that coordinates with the two active site aspartates, opening the door to structure-based drug design.[11] γ-secretase, however, has proven to be much more complicated. Nevertheless, the past few years have witnessed remarkable advances in our understanding of the properties, mechanism, and composition of this protease. This chapter will provide evidence that a multi-pass membrane protein called presenilin is the catalytic

Aβ Metabolism and Alzheimer's Disease, edited by Takaomi C. Saido. ©2003 Eurekah.com.

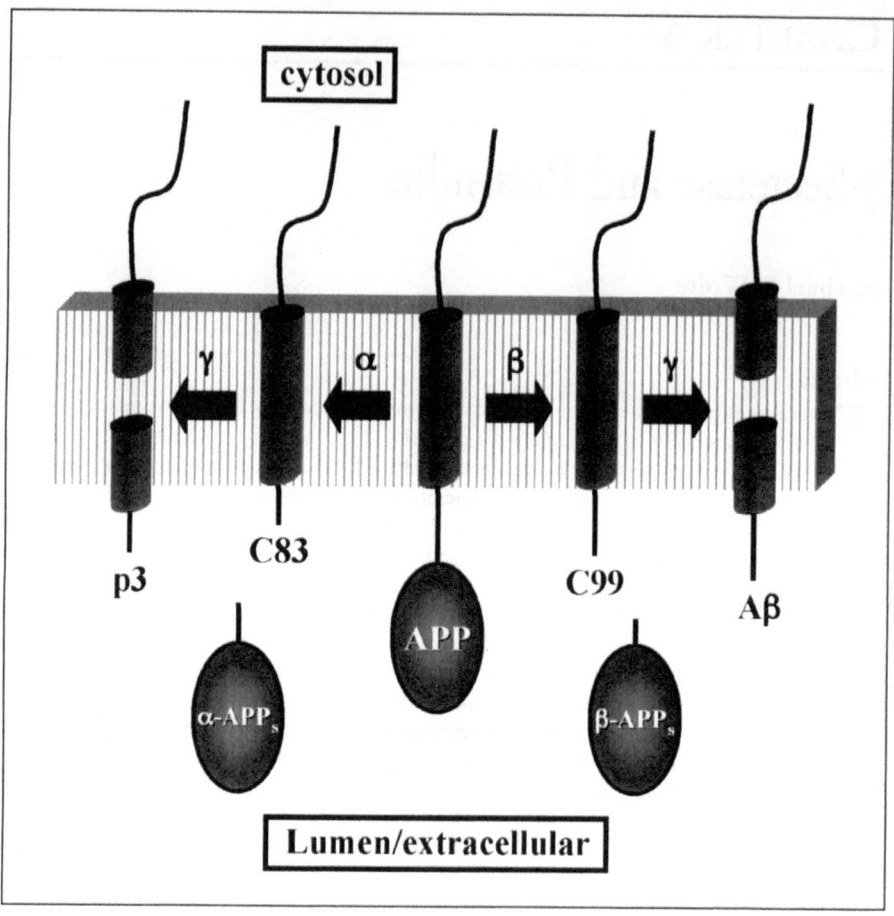

Figure 4.1. Topology and proteolytic processing of APP. Cleavage by α-secretase forms C83, while process-ing by β-secretase produces C99. In either event, the APP ectodomain is shed (α- and β-APP$_s$). C83 and C99 are further processed by γ-secretase within the transmembrane domain to produce p3 and Aβ, respectively.

component of a larger γ-secretase complex, address objections to the presenilin as protease hypothesis, place γ-secretase in context with other multi-pass membrane proteins, and project future directions on the study of this enzyme.

Substrate Specificity and Intramembrane Proteolysis

γ-secretase has been considered central to understanding the etiology of AD because it determines the proportion of the highly fibrillogenic 42 amino acid $Aβ_{42}$ peptide that is par-ticularly implicated in the pathogenesis of AD. After either α- or β-secretase release the APP ectodomain, the resulting C83 and C99 APP C-terminal fragments, are clipped in the middle their transmembrane regions by γ-secretase (Fig. 4.1).[12] Normally, about 90% of the proteoly-sis occurs between Val40 and Ile41 (Aβ numbering) to give the 40 amino acid $Aβ_{40}$, and roughly 10% takes place between Ala42 and Thr41 to produce $Aβ_{42}$ (Fig. 4.2). Minor propor-tions of other C-terminal variants, such as $Aβ_{39}$ and $Aβ_{43}$, are formed as well. γ-secretase has been of interest not only because of its key role in AD pathogenesis, but also because it presents

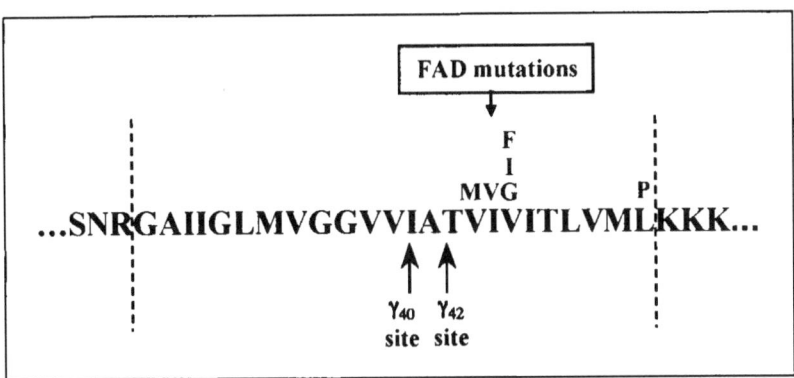

Figure 4.2. Sequence of APP in and around the γ-secretase cleavage site and disease-causing missense mutations.

an intriguing biochemical problem: how does this enzyme catalyze a hydrolysis at a site located within a membrane?

Does γ-secretase actually catalyze a hydrolysis within the boundaries of the lipid bilayer? Molecular modeling of the γ-secretase cleavage site in APP as an α-helix, a conformation typical of transmembrane domains, showed that the two major cleavage sites, one leading to $A\beta_{40}$ and the other to $A\beta_{42}$, are on opposite faces of the helix.[13] Moreover, in this model FAD-causing APP mutations are immediately adjacent to the scissile amide bond that leads to $A\beta_{42}$. Thus, the helical model provides a simple biochemical explanation for why these FAD mutations result in selective increases in $A\beta_{42}$ production. Lichtenthaler et al provided important experimental support for this model via a phenylalanine scanning mutagenesis study.[14] Systematic replacement of each APP transmembrane residue after the γ_{42} cut site with phenylalanine resulted in periodic changes in effects on $A\beta_{40}$ and $A\beta_{42}$ production that were consistent with a helical conformation of substrate upon initial interaction with the protease. In a second mutagenesis study, Lichtenthaler et al further showed that the site of γ-secretase cleavage is changed if the length of the transmembrane domain of APP is altered (e.g., by insertion or deletion of two hydrophobic residues on either end of the transmembrane domain). Thus, the length of the whole transmembrane domain is apparently a major determinant for the cleavage site of γ-secretase.[15] These findings are consistent with γ-secretase being a novel intramembrane-cleaving protease (I-CliP), which hydrolyzes its substrates within the confines of the membrane.[16]

Inhibitor Studies Suggest an Aspartyl Protease Mechanism

Important mechanistic evidence that γ-secretase is an aspartyl protease came from observations that hydroxyl-containing peptidomimetics inhibit γ-secretase activity[13,17,18] (Fig. 4.3). Such analogues mimic the *gem*-diol transition-state of aspartyl protease catalysis, and a variety of related compounds have been shown to inhibit other aspartyl proteases. For example, the general aspartyl protease inhibitor pepstatin A contains two hydroxyethyl moieties,[19] and the difluoro alcohol group is found in peptidomimetic inhibitors of renin[20] and penicillopepsin.[21] Potent substrate-based inhibitors of β-secretase contain the hydroxyethyl moiety.[22] Moreover, all of the clinically marketed inhibitors of HIV protease contain a hydroxyethyl group.[23] In many cases, cocrystal structures of these compounds and their cognate proteases have been determined (e.g., see refs. 11, 21,24), demonstrating that the hydroxyl group of the inhibitor indeed coordinates with the two active site aspartates in the protease.

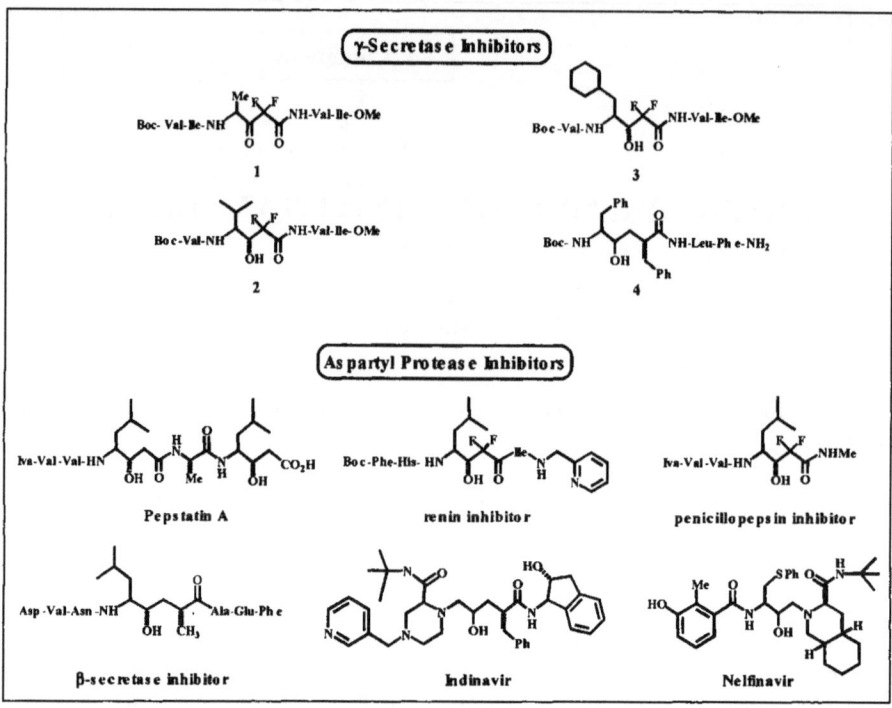

Figure 4.3. Transition-state inhibitors of γ-secretase and of known aspartyl proteases. Pepstatin A is a natural product that inhibits a wide spectrum of aspartyl proteases. Indinavir and nelfinavir exemplify currently marketed inhibitors of HIV protease.

The Presenilin/γ-Secretase Connection: AD-Causing Mutations and Knockout Mice

During the search for the major FAD-causing genes on chromosomes 14 and 1, many thought the encoded proteins would reveal at least one, if not both, of the proteases involved in Aβ production. When the search identified the presenilins in 1995,[25,26] it was far from clear what the normal function of these proteins might be and why mutant forms might lead to AD. The two proteins, presenilin-1 (PS1) and presenilin-2 (PS2), are 65% identical, but the only homology they had to anything else known at the time was to a set of otherwise obscure proteins in worms involved in egg-laying and spermatogenesis.[27,28] Remarkably, these proteins were the sites of dozens of FAD-causing missense mutations.[29] Over 70 such mutations have now been identified, with all but six occurring in PS1.[30] These mutations are found in many regions of the linear sequence, although they tend to cluster in certain areas (Fig. 4.4).

How could all these different PS mutations cause AD? Some mutations are extremely deleterious, a single copy of the mutant gene resulting in the onset of dementia as early as 25 years old.[31] Because of the amyloid plaque pathology of AD and because FAD-causing APP mutations are near β- and γ-secretase cleavage sites and affect Aβ production, investigating the effects of FAD-causing PS mutations on Aβ production, deposition, or toxicity seemed reasonable. Indeed, all PS mutations examined to date have been found to cause specific increases in Aβ$_{42}$ production, and this effect can be observed in transfected cells, in transgenic mice, and in blood plasma or media from cultured fibroblasts of FAD patients.[32-35] Thus, presenilins could somehow modulate γ-secretase activity to enhance cleavage of the Ala42-Thr43 amide bond. A

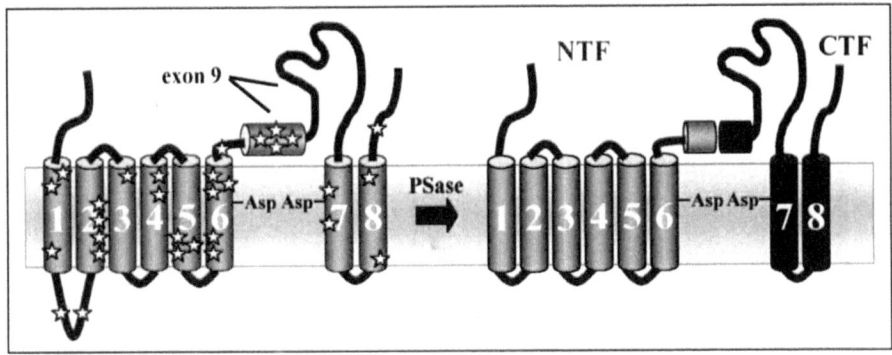

Figure 4.4. Topology and proteolytic processing of presenilins. Presenilins are cleaved within the hydrophobic region of the large cytosolic loop between TM6 and TM7, resulting in the formation of a heterodimeric complex composed of the N-terminal fragment (NTF) and the C-terminal fragment (CTF). Stars represent sites of missense mutation that cause familial Alzheimer's disease. The two conserved aspartates required for presenilin endoproteolysis and γ-secretase processing of APP and Notch are predicted to be within TM6 and TM7. The region of the protein encoded by exon 9 is denoted; natural deletion of this exon leads to a noncleavable but functional presenilin.

major clue to the function of the presenilins came via PS1 knockout mice: deletion of PS1 in these mice was lethal in utero, indicating the clear requirement of this gene for proper development of the organism.[36,37]

Unfortunately, the embryonic lethality resulting from deleting PS1 did not suggest reasons why PS mutations might cause AD relatively late in life. Nevertheless, neurons from PS1-deficient embryos could be cultured, and transfection of these cells with APP revealed that γ-secretase activity was markedly reduced.[38] The maturation and distribution of APP was not affected by the deletion of the PS1 gene, nor was the release of α- or β-APP$_s$ altered. However, γ-secretase substrates C83 and C99 were dramatically elevated, and Aβ production was substantially lowered. Formation of total Aβ and Aβ$_{42}$ was reduced to similar degrees (to roughly 20% of levels seen in fibroblasts from PS1 +/+ littermates), indicating that PS1 plays a role in the production of both Aβ$_{40}$ (which makes up 90% of all Aβ) and Aβ$_{42}$. The remaining γ-secretase activity was thought to be due to PS2. Indeed, the development of PS1/PS2 double knockout mice[39,40] allowed the culturing of embryonic stem cells, and transfection of APP demonstrated that complete absence of γ-secretase activity.[41,42] Thus, presenilins are absolutely required for the γ-secretase cleavage of APP.

Presenilin Topology and Maturation

Presenilins themselves undergo proteolytic processing within the hydrophobic region of the large cytosolic loop between transmembrane domain (TM) 6 and TM 7 (Fig. 4.4) to form stable heterodimeric complexes composed of the N- and C-terminal fragments.[43,44] These heterodimers are only produced to limited levels even upon overexpression of the holoprotein and may be found at the cell surface.[43,45-47] Expression of exogenous presenilins leads to replacement of endogenous presenilin heterodimers with the corresponding exogenous heterodimers, indicating competition for limiting cellular factors needed for stabilization and endoproteolysis.[48]

FAD-causing presenilin mutants are likewise processed to stable heterodimers with one exception, a missense mutation in PS1 that leads to the aberrant splicing out of exon 9, a region that encodes the endoproteolytic cleavage site.[43,45] This PS1 ΔE9 variant is an active presenilin, able to partially rescue a loss of function presenilin mutation in the worm C. elegans,[49,50] and like other FAD-causing presenilin mutants, causes increased $A\beta_{42}$ production.[50,51] Upon overexpression, most PS1 ΔE9 is rapidly degraded similar to unprocessed wild-type presenilins; however, a small portion of this PS1 variant is stabilized in cells[45,52] and forms a high molecular weight complex like the N-, and C-terminal fragments,[44,53] suggesting that it can interact with the same limiting cellular factors as wild-type presenilins. These observations are consistent with the idea that the bioactive form of presenilin is the heterodimer and that the hydrophobic region is an inhibitory domain.

Presenilins: Intramembrane-Cleaving Aspartyl Proteases

Presenilins contain two transmembrane aspartates (Fig, 4.4), one found in TM6 and one in TM7, predicted to lie the same distance within the membrane (i.e., they could interact with each other) and roughly aligned with the γ-secretase cleavage site in APP (i.e., they might work together to cut C99 and C83). These two aspartates are completely conserved from worms to humans and are even found in a plant presenilin.[54] Mutation of either TM aspartate to alanine did not affect the expression or subcellular location of APP, and the subcellular distribution of the mutant presenilins was also similar to the wild-type.[55] However, the mutant presenilins were completely incapable of undergoing endoproteolysis and acted in a dominant-negative manner with respect to γ-secretase processing of APP. Similar effects on APP processing were observed even when conservative mutations to glutamate[55] or asparagine[56,57] were made, indicating the crucial identity of these two key residues as aspartates and suggesting that the effects are not likely due to misfolding. These effects have been corroborated by several different laboratories and have been seen for both PS1 and PS2.[53,55-59]

The aspartates are critical for γ-secretase activity independent of their role in presenilin endoproteolysis: aspartate mutation in the PS1 ΔE9 variant still blocked γ-secretase activity, even though endoproteolysis is not required of this presenilin variant.[55] Together these results suggest that presenilins might be the catalytic component of γ-secretase: upon interaction with as yet unidentified limiting cellular factors, presenilin undergoes autoproteolysis via the two aspartates, and the two presenilin subunits remain together, each contributing one aspartate to the active site of γ-secretase. The issues of PS autoproteolysis and the role of PS endoproteolysis is controversial, especially in light of the identification of certain uncleavable artificial missense PS1 mutants are still functional with respect to γ-secretase activity.[56] On the other hand, these mutations may disrupt the putative pro domain so that it no longer blocks the active site.

Advancing the understanding of γ-secretase and the role of presenilins in this activity had been hampered by the lack of an isolated enzyme assay. Li et al reported a solubilized γ-secretase assay that faithfully reproduces the properties of the protease activity observed in whole cells.[60] Isolated microsomes were solubilized with detergent, and γ-secretase activity was determined by measuring Aβ production from a C-terminally modified version of C99. $A\beta_{40}$ and $A\beta_{42}$ were produced in the same ratio as seen in living cells (~9:1), and peptidomimetics that blocked $A\beta_{40}$ and $A\beta_{42}$ formation in cells likewise inhibited production of these Aβ species in the solubilized protease assay. Choice of detergent was critical for Aβ production in the assay: CHAPSO was optimal, although CHAPS, a detergent known to keep presenilin subunits together,[44] was also compatible with activity, and Triton-X100 did not allow any Aβ formation. After separation of the detergent-solubilized material by size-exclusion chromatography, γ-secretase activity coeluted with the two subunits of PS1. Remarkably, immunoprecipitated PS1 heterodimers also produced Aβ from the artificial substrate, strongly suggesting that presenilins are part of a large γ-secretase complex.

More direct evidence that presenilins are the catalytic components of γ-secretases came from affinity labeling studies using transition-state analogue inhibitors. Shearman et al identified a peptidomimetic γ-secretase inhibitor by rescreening compounds originally designed against HIV protease.[18] The compound blocks γ-secretase activity with an IC_{50} of 0.3 nM in the solubilized protease assay[60] and contains a hydroxyethyl isostere, a transition-state mimicking moiety found in many aspartyl protease inhibitors. While the transition-state mimicking alcohol directs the compound to aspartyl proteases, flanking substructures determine specificity. Indeed, this compound does not inhibit aspartyl proteases cathepsin D and HIV-1 protease.[18]

Photoactivatable versions of this compound bound covalently to presenilin subunits exclusively.[61] Interestingly, installation of the photoreactive group on one end of the inhibitor led to labeling of the N-terminal presenilin subunit, while installation on the other end resulted in the tagging of the C-terminal subunit. Moreover, while these agents did not label wild-type PS1 holoprotein, they did tag PS1 ΔE9, which as described above is not processed to heterodimers but nevertheless active. Similarly, Esler et al identified peptidomimetic inhibitors containing a difluoro alcohol group, another type of transition-state mimicking moiety, and these compounds were developed starting from a substrate-based inhibitor designed from the γ-secretase cleavage site in APP.[62] Conversion of one such analogue to a reactive bromoacetamide provided an affinity reagent that likewise bound covalently and specifically to PS1 subunits in cell lysates, isolated microsomes, and whole cells. Either PS1 subunit so labeled could be brought down with antibodies to the other subunit under coimmunoprecipitation conditions, demonstrating that the inhibitor bound to heterodimeric PS1. Seiffert and colleagues likewise identified presenilin subunits as the molecular target of novel peptidomimetic γ-secretase inhibitors. The affinity probe, however, does not resemble known transition-state mimics, so it is not clear whether this compound would be expected to bind to the active site of the protease.[63]

Taken together, these results strongly suggest that heterodimeric presenilin contains the catalytic component of γ-secretase: inhibitors in two of the three studies are transition-state analogues targeted to the active site. The wild-type presenilin holoprotein could be an inactive zymogen that requires cleavage into two subunits for activation. In any event, the active site is likely at the PS heterodimeric interface: both subunits are labeled by γ-secretase affinity reagents, and each contributes one critical aspartate. Whether a separate "presenilinase" converts the holoprotein to subunits or presenilins undergo autoproteolysis remains to be determined, although the absolute requirement of the two transmembrane aspartates for heterodimer formation suggests the latter.

Biological Roles of γ-Secretases

The presenilins are not only involved in the proteolytic processing of APP. They are also critical for processing of the Notch receptor, a signaling molecule crucial for cell-fate determination during embryogenesis.[64] After translation in the endoplasmic reticulum (ER), Notch is processed by a furin-like protease, resulting in a heterodimeric receptor that is shuttled to the cell surface[65] (Fig. 4.5). Upon interaction with a cognate ligand, the ectodomain of Notch is shed by a metalloprotease apparently identical to tumor necrosis factor-α converting enzyme (TACE).[66,67] Interestingly, metalloproteases such as TACE and ADAM-10 are among the identified α-secretases that shed the APP ectodomain[68,69] (see section *The Presenilin/γ-secretase Connection*). The membrane-associated C-terminus is then cut within the postulated transmembrane domain to release the Notch intracellular domain (NICD), which then translocates to the nucleus where it interacts with and activates the CSL family of transcription factors.[70] NICD formation is absolutely required for signaling from the Notch receptor: knock-in of a single point mutation near the transmembrane cleavage site in Notch1 results in an embryonic lethal phenotype in mice virtually identical to that observed upon knockout of the entire Notch1 protein.[71]

The parallels between APP and Notch processing are striking. Not only are both cleaved by TACE, but also the transmembrane regions of both proteins are processed by a γ-secretase-like protease that requires presenilins. Deletion of PS1 in mice is embryonic lethal, with a phenotype similar to that observed upon knockout of Notch1,[36,37] and the PS1/PS2 double knockout

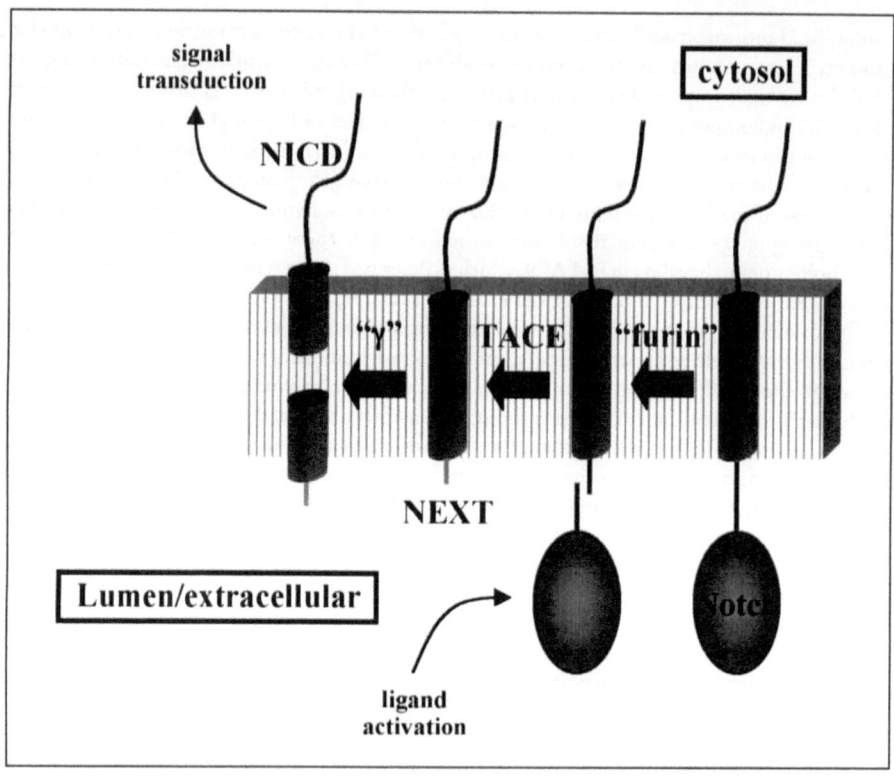

Figure 4.5. Topology and proteolytic processing of the Notch receptor. Notch is processed in the Golgi by a furin-like convertase and then transported to the cell surface as a heterodimer. Upon ligand binding, the ectodomain is shed by TACE, and the resulting membrane-associated fragment (Notch extracellular truncation; NEXT) is cleaved within its transmembrane domain by a γ-secretase-like protease, releasing the Notch intracellular domain (NICD).

phenotype is even more similar.[39,40] Deficiency in PS1 dramatically reduces NICD formation,[72] and the complete absence of presenilins results in total abolition of NICD production.[41,42] Treatment of cells with γ-secretase inhibitors designed from the transmembrane cleavage site within APP likewise blocks NICD production[72] and nuclear translocation[73] and reduces Notch signaling from a reporter gene.[73] Moreover, the two conserved TM aspartates in presenilins are required for cleavage of the Notch TM domain: as seen with γ-secretase inhibitors, expression of Asp-mutant PS1 or PS2 results in reduction of NICD formation, translocation and signaling.[59,73-75] Thus, if presenilins are the catalytic components of the γ-secretases that process APP, they are also likely the catalytic components of the related proteases that clip the transmembrane region of Notch.

Another γ-secretase substrate has recently been identified: the growth factor-dependent membrane tyrosine kinase receptor ErbB4.[76] Ectodomain shedding of ErbB4, but not other ErbB receptors, can be stimulated by activator of protein kinase C or binding with its cognate ligand heregulin. The remaining membrane-bound C-terminal fragment is cleaved to release the cytosolic domain in a PS-dependent process that is blocked by γ-secretase inhibitors. Inhibition of γ-secretase also prevented growth inhibition by heregulin. Thus, γ-secretase proteolysis may be part of a new mechanism for receptor tyrosine kinase signaling.

Yet another single-pass membrane protein called E-cadherin is also apparently a substrate for γ-secretase.[77] Cadherins are involved in cell-cell adhesion as well as in cell signaling, proliferation and differentiation. A presenilin-dependent γ-secretase-like protease mediates cleavage of E-cadherin at the membrane-cytosol interface: release of the cytosolic domain is dramatically reduced by either knockout of PS1 or treatment with a γ-secretase inhibitor. Both full-length E-cadherin as well as a C-terminal fragment resulting from ectodomain shedding by a metalloprotease are susceptible to this γ-secretase-like proteolysis. This stands in contrast to other γ-secretase substrates so far identified, which require ectodomain shedding for transmembrane processing. Both β- and α-catenins associate with the cytosolic tail of E-cadherin, and proteolysis of E-cadherin leads to release of these proteins into the cytosol. Such release may affect cytoskeletal architechure at points of cell-cell contact and β-catenin-mediated modulation of the Wnt signaling pathway. It seems likely that other substrates for γ-secretase will be identified, especially because γ-secretase lacks sequence specificity.

Objections to the Presenilin as Protease Hypothesis

While the above evidence strongly suggests that the catalytic component of γ-secretase resides in presenilin, this idea has not been without its skeptics.[78,79] Four apparent inconsistencies have been raised. First is the so-called "spatial paradox":[80] presenilin occurs mainly in the endoplasmic reticulum and Golgi apparatus, whereas the γ-secretase cleavages of Notch and APP are thought to occur at or near the cell surface. However, presenilin localization studies have used antibodies that do not distinguish between full-length and heterodimeric presenilin or between those heterodimers that have entered high-molecular weight, bioactive complexes and those that have not. Moreover, presenilin heterodimers have actually been detected in cell surface labeling experiments.[74] Second, Notch and APP are not processed in the same way: APP is cleaved heterogenously, whereas Notch is believed to be cut at a single site.[70] However, only one of the two Notch cleavage products, the cytosolic tail, has been analyzed. If only the cytosolic tail of APP had been characterized, then the conclusion would be that Notch and APP are cleaved at the same site.[81,82] Isolation and characterization of the Notch counterpart of Aβ may reveal heterogeneity as well. Third, mutating one of the two presenilin aspartates is reported to affect APP and Notch processing differently in cells.[75] However, this mutation does inhibit γ-secretase cleavage of both substrates, and the apparent differences are difficult to interpret with endogenous wild type presenilin in the background. Fourth, a class of γ-secretase inhibitors blocks the γ-secretase cleavage of APP but not Notch in cells.[83] These compounds, however, do not interact with γ-secretase directly in cell-free assays,[84] and their mechanism of action may be upstream of the enzyme.

Nicastrin and Other Putative Members of the γ-Secretase Complex

In 2000, Yu and colleagues identified another factor apparently critical for γ-secretase activity, a type I integral membrane protein they dubbed nicastrin.[85] This protein coimmunoprecipitated with anti-PS1 antibodies from glycerol gradient fractions containing high molecular weight (~250 KDa) presenilin complexes. RNA interference in *C. elegans* resulted in a Notch-deficient phenotype similar to that seen upon knockout of both worm presenilin genes. Human nicastrin associates with the γ-secretase substrates C83 and C99, and mutations in nicastrin can dramatically reduce Aβ production. More recently, knockout studies in *Drosophila* demonstrated that nicastrin is required for the γ-secretase processing of Notch and subsequent signaling.[86-88] Moreover, the development of an affinity purification method for γ-secretase using an immobilized transition-state analogue inhibitor led to the isolation of presenilin heterodimers and nicastrin.[89] Also, γ-secretase activity could be precipitated with anti-nicastrin antibodies. Thus, nicastrin appears to be an obligate member of the γ-secretase complex. The biochemical role nicastrin plays in substrate binding and turnover, however, is completely unknown.

Genetic studies in *C. elegans* and *D. melanogaster* have identified two other candidate members of the γ-secretase complex.[90,91] Loss-of-function mutations in both genes lead to developmental defects similar to those seen upon loss-of-function mutations in Notch or presenilin. Both genes, called aph-1 and pen-2, encode highly conserved multi-pass membrane proteins. Aph-1 is a ~30 kDa protein that contains seven putative transmembrane domains, and pen-2 is a ~10 kDa protein with two transmembrane domains. RNA interference against aph-1 or pen-2 results in dramatic reductions in Aβ production in *Drosophila* S2 cells transfected with human APP,[91] indicating that these two proteins are essential for γ-secretase activity. Biochemical studies are required to determine if these two integral membrane proteins are truly part of the protease complex along with presenilin and nicastrin. Recent work in our laboratory has indicated that human Aph-1 and Pen-2 do indeed form a complex with presenilin and nicastrin and that these proteins also bind to our γ-secretase affinity matrix in an activity-dependent manner. Overexpression of all four components (Aph-1, Pen-2, PS1, and nicastrin) results in dramatic increases in the levels of PS1 NTF, PS1 CTF, and a highly glycosylated form of nicastrin that tracks with γ-secretase activity. Overexpression of these proteins together also results in a dramatic elevation in γ-secretase activity. Our findings are consistent with the hypothesis that these four proteins assemble together, leading to PS endoproteolysis and the formation of active γ-secretase.

An Emerging Family of Polytopic Membrane Proteases

The hypothesis that presenilin is the catalytic component of γ-secretase has generated considerable controversy. The major reason for this is the lack of sequence homology with other known aspartyl proteases. However, another family of putative polytopic aspartyl proteases has been recently identified: type 4 prepilin peptidases.[92] These proteins also contain eight transmembrane domains and two conserved aspartates required for substrate proteolysis. However, in this case, the two aspartates do not appear to lie within transmembrane domains. Again, the type 4 prepilin peptidases do not bear sequence homology with other known (cytosolic) aspartyl proteases, but no other candidate protease has been proffered. More intriguing is the finding that signal peptide peptidase (SPP) is a seven-transmembrane protein that contains two conserved aspartate required for activity.[100] SPP shares signature sequence motifs with presenilins around the conserved aspartates, and expression of this protein in yeast reconstitutes protease activity. Thus, among the suspected members of the γ-secretase complex, only presenilin bears resemblance to a known protease, namely SPP.

In recent years, several other putative polytopic membrane proteases have been identified (see Table 4.1). These include:

The site 2 protease (S2P) that processes the sterol regulatory element binding protein (SREBP), a transcription factor essential for cholesterol biosynthesis.[93-95] An entire family of S2P-like proteins have been identified in bacteria and apparently process a mating factor.[96] The S2P family contains a conserved HEXXH motif essential to catalysis, suggesting that they are metalloproteases.

The yeast protein Ste24p that processes the CAAX C-terminus of certain proteins in conjunction with protein prenylation.[97] This protein has been purified to homogeneity and shown to cleave its substrate, demonstrating that Ste24p is a bona fide protease.[98] The requirement of zinc in this process suggests that Ste24p is a metalloprotease.

The rhomboid family of seven-TM proteins. Drosophila rhomboid-1 is responsible for the proteolysis of a membrane-anchored TGF-α-like protein called Spitz. Mutagenesis studies suggest that rhomboid-1 is a novel serine protease.[99]

Table 4.1. Polytopic membrane proteases

Mechanistic Class	Name	Substrates	I-CliP?
Metallo	S2P family	Ste24p, Rce1p	Yes
	SREBP, ATF6	CAAX prenylated proteins	No
Serine	Rhomboid family	EGF	Yes
Aspartic	Presenilins	APP, Notch, ErbB4, E-cadherin	Yes
	SPP family	Signal peptide remnants	Yes
	TFPP family	Leader peptides of Type 4 prepilins	No
Cysteine	?	?	?

Other than the identification of conserved and essential residues suggesting a particular proteolytic class, none of these proteins bears sequence homology with other known proteases. This appears to be an example of mechanistic convergence: nature has arrived at the same basic proteolytic mechanisms using polytopic proteins found in soluble proteins, suggesting that there are few other catalytic solutions to the problem of efficient proteolysis.

Future Directions

A considerable body of evidence now supports the hypothesis that the active site of γ-secretase lies at the interface of presenilin heterodimers and that γ-secretase is a complex of integral membrane proteins. Presenilins alone do not show proteolytic activity. Moreover, at the time of this writing, γ-secretase has not yet been purified to homogeneity nor has it been reconstituted into a background lacking this protease activity. Both of these goals, purification and reconstitution, will be required to identify all of the essential components of the γ-secretase complex. Only after this is accomplished can rigorous studies be carried out to determine the details of how this enzyme catalyzes hydrolysis within the hydrophobic environment of the lipid bilayer, how the protease complex is assembled, and how so many different mutations in presenilin result in increase $A\beta_{42}$ production. A more practical biomedical issue is whether this protease is an appropriate therapeutic target for AD. γ-secretase apparently processes a number of other proteins besides APP, and it is unclear whether chronic inhibition will be tolerated in aging adults. Resolution of this issue will require the development of potent, specific, and bioavailable inhibitors of this unusual, complex and important protease.

References

1. Chartier-Harlin MC, Crawford F, Houlden H et al. Early-onset Alzheimer's disease caused by mutations at codon 717 of the beta-amyloid precursor protein gene. Nature 1991; 353:844-6.
2. Goate A, Chartier-Harlin MC, Mullan M et al. Segregation of a missense mutation in the amyloid precursor protein gene with familial Alzheimer's disease. Nature 1991; 349:704-6.
3. Citron M, Oltersdorf T, Haass C et al. Mutation of the beta-amyloid precursor protein in familial Alzheimer's disease increases beta-protein production. Nature 1992; 360:672-4.
4. Cai XD, Golde TE & Younkin SG. Release of excess amyloid beta protein from a mutant amyloid beta protein precursor. Science 1993; 259:514-6.
5. Suzuki N, Cheung TT, Cai XD et al. An increased percentage of long amyloid beta protein secreted by familial amyloid beta protein precursor (beta APP717) mutants. Science 1994; 264:1336-40.
6. Haass C, Hung AY, Selkoe DJ et al. Mutations associated with a locus for familial Alzheimer's disease result in alternative processing of amyloid beta-protein precursor. J Biol Chem 1994; 269:17741-8.

7. Vassar R, Bennett BD, Babu-Khan S et al. beta-Secretase Cleavage of Alzheimer's Amyloid Precursor Protein by the Transmembrane Aspartic Protease BACE. Science 1999; 286:735-741.
8. Sinha S, Anderson JP, Barbour R et al. Purification and cloning of amyloid precursor protein beta-secretase from human brain. Nature 1999; 402:537-40.
9. Yan R, Bienkowski MJ, Shuck ME et al. Membrane-anchored aspartyl protease with Alzheimer's disease beta- secretase activity. Nature 1999; 402:533-7.
10. Hussain I, Powell D, Howlett DR et al. Identification of a Novel Aspartic Protease (Asp 2) as beta-Secretase. Mol Cell Neurosci 1999; 14:419-427.
11. Hong L, Koelsch G, Lin X et al. Structure of the protease domain of memapsin 2 (beta-secretase) complexed with inhibitor. Science 2000; 290:150-3.
12. Selkoe DJ. Cell biology of the amyloid beta-protein precursor and the mechanism of Alzheimer's disease. Annu Rev Cell Biol 1994; 10:373-403.
13. Wolfe MS, Xia W, Moore CL et al. Peptidomimetic probes and molecular modeling suggest Alzheimer's g-secretases are intramembrane-cleaving aspartyl proteases. Biochemistry 1999; 38:4720-7.
14. Lichtenthaler SF, Wang R, Grimm H et al. Mechanism of the cleavage specificity of Alzheimer's disease gamma- secretase identified by phenylalanine-scanning mutagenesis of the transmembrane domain of the amyloid precursor protein. Proc Natl Acad Sci USA 1999; 96:3053-3058.
15. Lichtenthaler SF, Beher D, Grimm HS et al. The intramembrane cleavage site of the amyloid precursor protein depends on the length of its transmembrane domain. Proc Natl Acad Sci USA 2002; 99:1365-70.
16. Wolfe MS, De Los Angeles J, Miller DD et al. Are presenilins intramembrane-cleaving proteases? Implications for the molecular mechanism of Alzheimer's disease. Biochemistry 1999; 38:11223-30.
17. Moore CL, Leatherwood DD, Diehl TS et al. Difluoro Ketone Peptidomimetics Suggest a Large S1 Pocket for Alzheimer's γ-secretase: Implications for Inhibitor Design. J Med Chem 2000; 43:3434-3442.
18. Shearman MS, Beher D, Clarke EE et al. L-685,458, an Aspartyl Protease Transition State Mimic, Is a Potent Inhibitor of Amyloid beta-Protein Precursor gamma-Secretase Activity. Biochemistry 2000; 39:8698-8704.
19. Marciniszyn J Jr, Hartsuck JA, Tang J. Mode of inhibition of acid proteases by pepstatin. J Biol Chem 1976; 251:7088-94.
20. Thaisrivongs S, Pals DT, Kati WM et al. Design and synthesis of potent and specific renin inhibitors containing difluorostatine, difluorostatone, and related analogues. J Med Chem 1986; 29:2080-7.
21. James MN, Sielecki AR, Hayakawa K et al. Crystallographic analysis of transition state mimics bound to penicillopepsin: difluorostatine- and difluorostatone-containing peptides. Biochemistry 1992; 31:3872-86.
22. Ghosh AK, Shin D, Downs D et al. Design of potent inhibitors for human brain memapsin 2 (beta-secretase). J Am Chem Soc 2000; 122:3522-3.
23. Huff JR. HIV protease: a novel chemotherapeutic target for AIDS. J Med Chem 1991; 34:2305-14.
24. Silva AM, Cachau RE, Sham HL et al. Inhibition and catalytic mechanism of HIV-1 aspartic protease. J Mol Biol 1996; 255:321-46.
25. Sherrington R, Rogaev EI, Liang Y et al. Cloning of a gene bearing missense mutations in early-onset familial Alzheimer's disease. Nature 1995; 375:754-60.
26. Levy-Lahad E, Wasco W, Poorkaj P et al. Candidate gene for the chromosome 1 familial Alzheimer's disease locus. Science 1995; 269:973-7.
27. L'Hernault SW, Arduengo PM. Mutation of a putative sperm membrane protein in Caenorhabditis elegans prevents sperm differentiation but not its associated meiotic divisions. J Cell Biol 1992; 119:55-68.
28. Levitan D, Greenwald I. Facilitation of lin-12-mediated signalling by sel-12, a Caenorhabditis elegans S182 Alzheimer's disease gene. Nature 1995; 377:351-4.
29. Hardy J. Amyloid, the presenilins and Alzheimer's disease. Trends Neurosci 1997; 20:154-9.
30. http://www.alzforum.org/res/com/mut/pre/default.asp.
31. Campion D, Dumanchin C, Hannequin D et al. Early-onset autosomal dominant Alzheimer disease: prevalence, genetic heterogeneity, and mutation spectrum. Am J Hum Genet 1999; 65:664-70.

32. Scheuner D, Eckman C, Jensen M et al. Secreted amyloid beta-protein similar to that in the senile plaques of Alzheimer's disease is increased in vivo by the presenilin 1 and 2 and APP mutations linked to familial Alzheimer's disease. Nat Med 1996; 2:864-70.

33. Citron M, Westaway D, Xia W et al. Mutant presenilins of Alzheimer's disease increase production of 42-residue amyloid beta-protein in both transfected cells and transgenic mice. Nat Med 1997; 3:67-72.

34. Tomita T, Maruyama K, Saido TC et al. The presenilin 2 mutation (N141I) linked to familial Alzheimer disease (Volga German families) increases the secretion of amyloid beta protein ending at the 42nd (or 43rd) residue. Proc Natl Acad Sci USA 1997; 94:2025-30.

35. Duff K, Eckman C, Zehr C et al. Increased amyloid-beta42(43) in brains of mice expressing mutant presenilin 1. Nature 1996; 383:710-3.

36. Wong PC, Zheng H, Chen H et al. Presenilin 1 is required for Notch1 and Dll1 expression in the paraxial mesoderm. Nature 1997; 387:288-92.

37. Shen J, Bronson RT, Chen DF et al. Skeletal and CNS defects in Presenilin-1-deficient mice. Cell 1997; 89:629-39.

38. De Strooper B, Saftig P, Craessaerts K et al. Deficiency of presenilin-1 inhibits the normal cleavage of amyloid precursor protein. Nature 1998; 391:387-90.

39. Donoviel DB, Hadjantonakis AK, Ikeda M et al. Mice lacking both presenilin genes exhibit early embryonic patterning defects. Genes Dev 1999; 13:2801-10.

40. Herreman A, Hartmann D, Annaert W et al. Presenilin 2 deficiency causes a mild pulmonary phenotype and no changes in amyloid precursor protein processing but enhances the embryonic lethal phenotype of presenilin 1 deficiency. Proc Natl Acad Sci USA 1999; 96:11872-7.

41. Herreman A, Serneels L, Annaert W et al. Total inactivation of gamma-secretase activity in presenilin-deficient embryonic stem cells. Nat Cell Biol 2000; 2:461-462.

42. Zhang Z, Nadeau P, Song W et al. Presenilins are required for gamma-secretase cleavage of beta-APP and transmembrane cleavage of Notch-1. Nat Cell Biol 2000; 2:463-465.

43. Thinakaran G, Borchelt DR, Lee MK et al. Endoproteolysis of presenilin 1 and accumulation of processed derivatives in vivo. Neuron 1996; 17:181-90.

44. Capell A, Grunberg J, Pesold B et al. The proteolytic fragments of the Alzheimer's disease-associated presenilin-1 form heterodimers and occur as a 100-150-kDa molecular mass complex. J Biol Chem 1998; 273:3205-11.

45. Ratovitski T, Slunt HH, Thinakaran G et al. Endoproteolytic processing and stabilization of wild-type and mutant presenilin. J Biol Chem 1997; 272:24536-41.

46. Podlisny MB, Citron M, Amarante P et al. Presenilin proteins undergo heterogeneous endoproteolysis between Thr291 and Ala299 and occur as stable N- and C-terminal fragments in normal and Alzheimer brain tissue. Neurobiol Dis 1997; 3:325-37.

47. Steiner H, Capell A, Pesold B et al. Expression of Alzheimer's disease-associated presenilin-1 is controlled by proteolytic degradation and complex formation. J Biol Chem 1998; 273:32322-31.

48. Thinakaran G, Harris CL, Ratovitski T et al. Evidence that levels of presenilins (PS1 and PS2) are coordinately regulated by competition for limiting cellular factors. J Biol Chem 1997; 272:28415-22.

49. Levitan D, Doyle TG, Brousseau D et al. Assessment of normal and mutant human presenilin function in Caenorhabditis elegans. Proc Natl Acad Sci USA 1996; 93:14940-4.

50. Steiner H, Romig H, Grim MG et al. The biological and pathological function of the presenilin-1 Dexon 9 mutation is independent of its defect to undergo proteolytic processing. J Biol Chem 1999; 274:7615-8.

51. Borchelt DR, Thinakaran G, Eckman CB et al. Familial Alzheimer's disease-linked presenilin 1 variants elevate Abeta1-42/1-40 ratio in vitro and in vivo. Neuron 1996; 17:1005-13.

52. Zhang J, Kang DE, Xia W et al. Subcellular distribution and turnover of presenilins in transfected cells. J Biol Chem 1998; 273:12436-42.

53. Yu G, Chen F, Nishimura M et al. Mutation of conserved aspartates affect maturation of both aspartate-mutant and endogenous presenilin 1 and presenilin 2 complexes. J Biol Chem 2000; 275:27348-53.

54. Lin X, Kaul S, Rounsley S et al. Sequence and analysis of chromosome 2 of the plant Arabidopsis thaliana [see comments]. Nature 1999; 402:761-8.

55. Wolfe MS, Xia W, Ostaszewski BL et al. Two transmembrane aspartates in presenilin-1 required for presenilin endoproteolysis and γ-secretase activity. Nature 1999; 398:513-7.
56. Steiner H, Romig H, Pesold B et al. Amyloidogenic function of the Alzheimer's disease-associated presenilin 1 in the absence of endoproteolysis. Biochemistry 1999; 38:14600-5.
57. Leimer U, Lun K, Romig H et al. Zebrafish (Danio rerio) presenilin promotes aberrant amyloid beta- peptide production and requires a critical aspartate residue for its function in amyloidogenesis. Biochemistry 1999; 38:13602-9.
58. Kimberly WT, Xia W, Rahmati T et al. The transmembrane aspartates in presenilin 1 and 2 are obligatory for γ-secretase activity and amyloid β-protein generation. J Biol Chem 2000; 275:3173-8.
59. Steiner H, Duff K, Capell A et al. A loss of function mutation of presenilin-2 interferes with amyloid beta-peptide production and notch signaling. J Biol Chem 1999; 274:28669-73.
60. Li YM, Lai MT, Xu M et al. Presenilin 1 is linked with gamma -secretase activity in the detergent solubilized state. Proc Natl Acad Sci USA 2000; 97:6138-6143.
61. Li YM, Xu M, Lai MT et al. Photoactivated gamma-secretase inhibitors directed to the active site covalently label presenilin 1. Nature 2000; 405:689-94.
62. Esler WP, Kimberly WT, Ostaszewski BL et al. Transition-state analogue inhibitors of γ-secretase bind directly to presenilin-1. Nature Cell Biology 2000; 2:428-34.
63. Seiffert D, Bradley JD, Rominger CM et al. Presenilin-1 and -2 are molecular targets for gamma -secretase inhibitors. J Biol Chem 2000; 275:34086-91.
64. Artavanis-Tsakonas S, Rand MD & Lake RJ. Notch signaling: cell fate control and signal integration in development. Science 1999; 284:770-6.
65. Logeat F, Bessia C, Brou C et al. The Notch1 receptor is cleaved constitutively by a furin-like convertase. Proc Natl Acad Sci USA 1998; 95:8108-12.
66. Brou C, Logeat F, Gupta N et al. A Novel Proteolytic Cleavage Involved in Notch Signaling: The Role of the Disintegrin-Metalloprotease TACE. Molecular Cell 2000; 5:207-216.
67. Mumm JS, Schroeter EH, Saxena MT et al. A Ligand-Induced Extracellular Cleavage Regulates -Secretase-like Proteolytic Activation of Notch1. Molecular Cell 2000; 5:197-206.
68. Buxbaum JD, Liu KN, Luo Y et al. Evidence that tumor necrosis factor a converting enzyme is involved in regulated a-secretase cleavage of the Alzheimer's amyloid protein precursor. J. Biol. Chem. 1998; 273:27765-27767.
69. Lammich S, Kojro E, Postina R et al. Constitutive and regulated alpha-secretase cleavage of Alzheimer's amyloid precursor protein by a disintegrin metalloprotease. Proc Natl Acad Sci USA 1999; 96:3922-7.
70. Schroeter EH, Kisslinger JA, Kopan R. Notch-1 signalling requires ligand-induced proteolytic release of intracellular domain. Nature 1998; 393:382-6.
71. Huppert SS, Le A, Schroeter EH et al. Embryonic lethality in mice homozygous for a processing-deficient allele of Notch1. Nature 2000; 405:966-70.
72. De Strooper B, Annaert W, Cupers P et al. A presenilin-1-dependent γ-secretase-like protease mediates release of Notch intracellular domain. Nature 1999; 398:518-22.
73. Berezovska O, Jack C, McLean P et al. Aspartate mutations in presenilin and γ-secretase inhibitors both impair Notch1 proteolysis and nuclear translocation with relative preservation of Notch1 signaling. J Neurochem 2000; 75:583-593.
74. Ray WJ, Yao M, Mumm J et al. Cell surface presenilin-1 participates in the γ-secretase-like proteolysis of Notch. J Biol Chem 1999; 274:36801-7.
75. Capell A, Steiner H, Romig H et al. Presenilin-1 differentially facilitates endoproteolysis of the beta-amyloid precursor protein and Notch. Nat Cell Biol 2000; 2:205-11.
76. Ni CY, Murphy MP, Golde TE et al. gamma -Secretase cleavage and nuclear localization of ErbB-4 receptor tyrosine kinase. Science 2001; 294:2179-81.
77. Marambaud P, Shioi J, Serban G et al. A presenilin-1/gamma-secretase cleavage releases the E-cadherin intracellular domain and regulates disassembly of adherens junctions. Embo J 2002; 21:1948-56.
78. Sisodia SS, St George-Hyslop PH. gamma-Secretase, Notch, Abeta and Alzheimer's disease: where do the presenilins fit in? Nat Rev Neurosci 2002; 3:281-90.
79. Sisodia SS, Annaert W, Kim SH et al. Gamma-secretase: never more enigmatic. Trends Neurosci 2001; 24:S2-6.

80. Annaert WG, Levesque L, Craessaerts K et al. Presenilin 1 controls gamma-secretase processing of amyloid precursor protein in pre-golgi compartments of hippocampal neurons. J Cell Biol 1999; 147:277-94.

81. Sastre M, Steiner H, Fuchs K et al. Presenilin-dependent gamma-secretase processing of beta-amyloid precursor protein at a site corresponding to the S3 cleavage of Notch. EMBO Rep 2001; 2:835-41.

82. Yu C, Kim SH, Ikeuchi T et al. Characterization of a presenilin-mediated amyloid precursor protein carboxyl-terminal fragment gamma. Evidence for distinct mechanisms involved in gamma-secretase processing of the APP and Notch1 transmembrane domains. J Biol Chem 2001; 276:43756-60.

83. Petit A, Bihel F, Alves da Costa C et al. New protease inhibitors prevent gamma-secretase-mediated production of Abeta40/42 without affecting Notch cleavage. Nat Cell Biol 2001; 3:507-11.

84. Esler WP, Das C, Campbell WA et al. Amyloid-lowering isocoumarins are not direct inhibitors of γ-secretase. Nature Cell Biology 2002; 4:E110-1.

85. Yu G, Nishimura M, Arawaka S et al. Nicastrin modulates presenilin-mediated notch/glp-1 signal transduction and betaAPP processing. Nature 2000; 407:48-54.

86. Chung HM, Struhl G. Nicastrin is required for Presenilin-mediated transmembrane cleavage in Drosophila. Nat Cell Biol 2001; 3:1129-32.

87. Hu Y, Ye Y, Fortini ME. Nicastrin is required for gamma-secretase cleavage of the Drosophila Notch receptor. Dev Cell 2002; 2:69-78.

88. Lopez-Schier H, St Johnston D. Drosophila nicastrin is essential for the intramembranous cleavage of notch. Dev Cell 2002; 2:79-89.

89. Esler WP, Kimberly WT, Ostaszewski BL et al. Activity-dependent isolation of the presenilin/ γ-secretase complex reveals nicastrin and a g substrate. Proc Natl Acad Sci USA 2002; 99:2720-2725.

90. Goutte C, Tsunozaki M, Hale VA et al. APH-1 is a multipass membrane protein essential for the Notch signaling pathway in Caenorhabditis elegans embryos. Proc Natl Acad Sci USA 2002; 99:775-9.

91. Francis R, McGrath G, Zhang J et al. aph-1 and pen-2 Are Required for Notch Pathway Signaling, gamma-Secretase Cleavage of betaAPP, and Presenilin Protein Accumulation. Dev Cell 2002; 3:85-97.

92. LaPointe CF, Taylor RK. The type 4 prepilin peptidases comprise a novel family of aspartic acid proteases. J Biol Chem 2000; 275:1502-10.

93. Rawson RB, Zelenski NG, Nijhawan D et al. Complementation cloning of S2P, a gene encoding a putative metalloprotease required for intramembrane cleavage of SREBPs. Mol Cell 1997; 1:47-57.

94. Duncan EA, Dave UP, Sakai J et al. Second-site cleavage in sterol regulatory element-binding protein occurs at transmembrane junction as determined by cysteine panning. J Biol Chem 1998; 273:17801-9.

95. Ye J, Dave UP, Grishin NV et al. Asparagine-proline sequence within membrane-spanning segment of SREBP triggers intramembrane cleavage by site-2 protease. Proc Natl Acad Sci USA 2000; 97:5123-8.

96. Rudner DZ, Fawcett P, Losick R. A family of membrane-embedded metalloproteases involved in regulated proteolysis of membrane-associated transcription factors. Proc Natl Acad Sci USA 1999; 96:14765-70.

97. Boyartchuk VL, Ashby MN, Rine J. Modulation of Ras and a-factor function by carboxyl-terminal proteolysis. Science 1997; 275:1796-800.

98. Tam A, Schmidt WK & Michaelis S. The multispanning membrane protein Ste24p catalyzes CAAX proteolysis and NH2-terminal processing of the yeast α-factor precursor. J Biol Chem 2001; 276:46798-806.

99. Urban S, Lee JR & Freeman M. Drosophila rhomboid-1 defines a family of putative intramembrane serine proteases. Cell 2001; 107:173-82.

100. Weihofen A, Binns K, Lemberg MK et al. Identification of signal peptide peptidase, a presenilin-type aspartyl protease. Science 2002; 296:2215-8.

CHAPTER 5

Functional Roles of APP Secretases

Dieter Hartmann

Abstract

Cleavage of APP by α-, β- and γ-secretases strikingly resembles regulated intramembrane proteolysis (RIP), which is normally employed to generate signal-transducing fragments from transmembrane proteins. Protein 'RIPping' is initiated by signals like ligand binding causing the removal of the ectodomain by a sheddase-like activity. This exposes the transmembrane stub to an intramembrane-cleaving protease (i-Clip) releasing the intracellular domain that acts as a signal transducer directed towards the nucleus. At present, the presumed signaling function of APP remains largely enigmatic, as both the putative signal triggering RIPping and the elicited cellular response in vertebrates are unknown. Signaling deficits have mostly been defined in *C. elegans*, inhibiting pharyngeal motility.

However, further analysis of the protease activities involved in APP RIPping has shown that these enzymes also mediate shedding or RIPping of other proteins, some of which are of critical importance for development and maintenance of the organism. Disintegrin metalloproteases ADAM 9, 10 and 17, which together appear to constitute α-secretase, are sheddases for EGF-R ligands (e.g., TGF-α) as well as other proteins like TNF-α, L1 and L-selectin. ADAMs 10 and 17 also initiate the RIPping of Notch receptors, which are critical for cell fate selection. Likewise, γ-secretase-like activities depending on presenilins 1 and 2 are i-Clips for several type1 transmembrane proteins like APLPs, Notch receptors, erbB4 and possibly ATF6. Via a nonenzymatic binding of their loop domain, presenilins regulate cytosolic β-catenin concentration and thereby critically modulate the Wnt signaling pathway. Recent data also indicate a direct participation of presenilins in cadherin-mediated cell adhesion. Only for the β-secretase BACE1, such additional functions appear to be rare. So far, only the shedding of an otherwise Golgi-resident sialyl transferase was described and appears to be of low functional significance.

The surprising finding that a small number of proteases forms a kind of common shedding and RIPping machinery for a growing number of proteins not only underscores the functional importance of these proteases (more often than not illustrated by disastrous knockout phenotypes) but also indicates that the pharmacological targeting of these enzymes to treat Alzheimer's disease will have to be very carefully evaluated to avoid severe side effects.

Introduction: Cleavage of APP is a Special Case of Regulated Intramembrane Proteolysis

The cleavage of amyloid precursor protein (APP) by three different protease activities termed α-, β- and γ-secretase is the decisive event by which either the pathogenic amyloid β (Aβ) peptides or the harmless (and possibly even beneficial) APPsα fragments are released.[1,2]

Thereby, APP processing adheres to a stereotyped pattern of ectodomain secretion by either α- or β-secretase cleaving shortly outside to the cell surface followed by intramembrane proteolysis of the C-terminal stub by γ-secretase. The regulated ectodomain shedding—in other cases triggered by specific signals like ligand binding,[3-5] changing concentrations of a metabolite[6,7] or ER 'stress'[8]—renders the transmembrane fragment accessible to intramembrane proteolysis by an intramembrane-cleaving protease (i-Clip) releasing a signal-transducing intracellular fragment acting upon gene expression. This pattern of regulated intramembrane proteolysis (RIP) is evolutionarily conserved in a broad spectrum of organisms, ranging from mammals "down" to bacteria.[9] In parallel, "isolated" shedding of protein ectodomains has been identified as a part of the signaling mechanism in a number of growth factors like TNF-α and TGF-α or HB-EGF.[10-13]

As a peculiarity of APP, ectodomain shedding is performed by two alternative protease activities, α-secretase or β-secretase, whereby both 'versions' of ectodomain loss appear to result in the same signal-transducing fragment. As yet, the hypothetical signal triggering APP 'RIPping'—if such a signal exists at all—remains to be identified. Likewise, knowledge about the functional significance of this APP signaling is at best scarce.

APP is a member of a protein family for which as yet two other members, APP-like protein 1 (APLP1) and APLP2, have been identified. They are processed by the same secretase activities eventually releasing corresponding intracellular domains. A number of diverse functions have been—more or less speculatively—discussed for APP and APLPs during the past years, including the function as a receptor protein, cell adhesion molecule or growth factor (precursor).

The 'traditional' approach to define gene function by the analysis of knockout models, in this case also including APP/APLP single and double knockouts to address compensation effects, has yielded surprisingly uninformative results.[14,15] Mice deficient for APP or APLP1 develop almost normally except for minor growth retardation, ataxia and unspecific symptoms of brain pathology like astrogliosis. Likewise, APLP2 deficient mice are viable and only double-deficient APP/APLP2 or APLP1/APLP2 mice display a high mortality in the first post-natal week. The surviving ones show a considerable growth deficit and a panel of neurological abnormalities, which has however not been linked so far to any specific central nervous system (CNS) lesion. This absence of any specific identifiable defect explaining the severe double knockout phenotypes is at present the most severe obstacle in this important line of research.

The putative APP signaling function depending on intact secretase activity is likewise enigmatic. Intramembrane proteolysis generates the APP intracellular domain (AICD). AICD is able to bind to the adaptor protein Fe65 and to then enter the nucleus activating gene transcription via binding of Fe65 to the transcription factor CP2/LPF1.[16,17] Additionally, Fe65 may bind to the C-terminus of APP in focal adhesion complexes, where it influences cell motility.[18] Possibly comparable results have been obtained in *C. elegans*, where the inactivation of either the homologues of APP (apl-1) or Fe65 (feh-1) has the identical inhibiting effect on pharyngeal pumping motility.[19] Another possibly important interaction has been described with mdab1, linking APP to the reelin pathway involved in the development of cortical cell layering.[20]

In contrast to our scant knowledge about the function of APP cleavage, considerable progress has been made in the identification of the responsible proteases, which are important drug targets for the therapy of AD.[21] Thus, α-secretase activity is apparently distributed over several enzymes, the best documented candidates of which are three disintegrin metalloproteases termed ADAM 9 / meltrin γ, [21] ADAM 10 / kuzbanian[23-25] and ADAM 17 / TACE.[26]

β-secretase activity is linked to a novel membrane-bound aspartyl protease termed BACE 1 / memapsin 2 / Asp2.[27-30] A related protein, BACE 2 /Asp1[31-34] more efficiently exerts a variant form of α-cleavage, its cleavage site being shifted by two amino acids in comparison to the 'original' α site.

γ-secretase activity is bound to a large multimeric complex centered around presenilin (PS) 1 and 2,[35-40] Within this complex, the actual protease activity is probably exerted by the PSs themselves, which could act as atypical aspartyl proteases.

The APP-processing secretase machinery appears to be astonishingly standardized in the sense that the same sets of sheddases and iCLIPs act upon several substrates, some of which are of critical importance for development and integrity of an organism. As a result, an impairment or loss of function of these secretases in many cases causes severe defects unrelated to APP/APLP knockout phenotypes.[12,41-46]

α-Secretase

Of the many different proteases proposed during the past years as α-secretase candidates, the discussion is now focusing on ADAM-type metalloproteases. ADAMs are type1 transmembrane proteins with a highly modular molecular architecture, strikingly similar to snake venom reprolysins.[47] They consist of a N-terminal prodomain inhibiting enzymatic activity, a protease domain, and disintegrin and cystein-rich domains involved in cell adhesion and fusion, a transmembrane domain and a short cytoplasmic tail. In mammals, this family consists of >30 members, of which only a minority appears to have retained a functional, active proteolytic site. For both proteolytic and adhesive functions deduced from the architecture of ADAM proteins, important 'applications' in animal biology have been identified. Thus, ADAM 2 mediates sperm/egg interaction and ADAM 12 / meltrin α myoblast fusion.[48] Proteolytically speaking, several ADAMs are important sheddases and initiators of protein RIPping with a partially overlapping substrate specificity, but may also act on extracellular matrix during cell migration.[49,50]

According to this overall profile, the three ADAMs in question here are both implied in 'simple' ectodomain shedding of a variety of proteins (for instance cleavage of HB-EGF by ADAM 9 or shedding of TGF-α and TNF-α in the case of ADAM 17) as well as in the initiation of protein RIPping as in the case of Notch receptor ectodomain ('S 2' cleavage) by ADAM 10 and ADAM 17. Many currently identified cases of protein shedding might turn out to be further examples for protein RIPping, as has been the case with erbB4 in the last year.[51]

ADAM 9

Expression of ADAM 9 is ubiquitous during development, with especially high levels in brain, heart and mesenchyme.[52] For adult tissues, presence of ADAM 9 has been reported for brain, kidney, bone (both osteoblasts and osteoclasts),[53] and endometrium. In kidney, ADAM 9 localizes to the basolateral membrane of tubular epithelial cells and visceral glomerular epithelium. Its colocalization with beta1 integrin chains has led to the assumption of an involvement in cell-cell and cell-basal lamina contacts.[54]

In endometrium, ADAM 9 undergoes a phasic regulation with a peak expression during the periimplantation period. It is further upregulated at the implantation site by the contacting blastocyst and may be involved in cell fusion events at the nidation site.[55] A participation of ADAM 9 in cell fusion has also been reported during giant cell formation from blood monocytes. Cell fusion in vitro at least in part depends on protease activity, as it could both be inhibited by antibodies against ADAM 9 and by the metalloprotease inhibitor SI 27 (Namba et al, 2001).

Additional functional data have been obtained from biochemical studies. Upon stimulation with TPA, ADAM 9 releases the ectodomain of HB-EGF into the culture medium via a mechanism involving binding of PKC δ to its cytoplasmic domain (Izumi et al, 1998). Furthermore, the cytoplasmic domain of ADAM 9 interacts with MAD2beta,[56] a protein related to the cell cycle regulator MAD2. However, a connection of ADAM 9 to cell cycle regulation is as yet not proven or further characterized by any functional assay.

In contrast to these reports pointing to the involvement of ADAM 9 in a number of important cell functions, the recently published ADAM 9 knockout mouse[52] is asymptomatic and fertile. Moreover, the secretion of both APP and HB-EGF was unaltered, indicating that the reported contribution of ADAM 9 to both events is either quantitatively unimportant or that its loss is readily replaced by other candidate α-secretases in the case of APP or by e.g., ADAM 12 in the case of HB-EGF.[57] However, several functions of ADAM 9 might be of special relevance predominantly under pathologic conditions, for instance its role in giant cell formation and neointima formation during vascular occlusion. Thus, the current characterization of this knockout model urgently awaits completion by tests challenging e.g., monocyte / macrophage function.[52]

ADAM 10

In contrast to ADAMs 9 and 17, which have exclusively been analyzed in vertebrates, orthologues of ADAM 10 have been identified early in *Drosophila* (termed *kuzbanian*),[41,58,59] and *C. elegans* (termed *sup-17*),[97,42] As a consequence, a major part of the genetic and functional analysis of ADAM 10 was performed in these species, and currently many of the data still have to be transferred to mammals. Already, these attempts have led to a number of conflicting results, partially being due to the fact that the identification of related proteases (e.g., orthologues of ADAMs 9 and 17) with possibly overlapping substrate specificities in both invertebrates is still incomplete.

The key "non-APP" function of ADAM 10 orthologues in both Drosphila and *C. elegans* is its involvement in the Notch pathway. By genetic evidence, it was shown that kuzbanian and SUP-17 are essential for Notch signaling and act in a cell-autonomous fashion. Their loss of function in both invertebrate models resulted in phenotypes closely related to the inactivation of Notch receptors.[42,58-60]

In parallel, biochemical studies have shown that triggered by ligand binding Notch receptors undergo ectodomain shedding ("S2"-cleavage) by an α-secretase-like activity as a prerequisite to the γ-secretase-mediated release of the signal-transducing NICD fragment ("S3-cleavage"). According to this model, loss of ADAM 10/kuzbanian/sup-17 should result in a phenotype virtually identical to a total Notch receptor inactivation (e.g., a Notch knockout in Drosophila or a hypothetical quadruple notch1-4 knockout in mammals). As also evidence has been published for a secretase-independent Notch signaling pathway, a more precise correlation might be drawn between a loss of (ADAM 10 mediated) S2 cleavage and γ-secretase-mediated S3-cleavage, both interrupting the protease-dependant Notch signaling pathway at two linearly consecutive steps.

Whereas this correlation between a Notch and/or PS loss-of-function and inactivation of the respective ADAM 10 orthologues is fairly precise in Drosophila and *C. elegans* (up to the generation of "Notch-like" defects in Xenopus expressing dominant-negative Drosophila kuzbanian variants), analysis of ADAM 10—negative murine fibroblasts surprisingly did not show effects on Notch receptor processing.[3,5] This stands in striking contrast to the phenotype of ADAM 10 deficient mice, which closely resembles that of PS double knockout embryos (Hartmann et al, in press). Such an impairment of Notch site 2 cleavage was however reported for ADAM 17 knockout fibroblasts, in contradiction to the fact that ADAM 17 deficient mice do not exhibit a phenotype consistent with defective Notch signaling.[3,5,12] For Drosophila, an additional impact of ADAM 10 / kuzbanian on Notch signaling could result from its role in the shedding of the Notch ligand delta (Qi et al, 1999), the biological significance of which is however unclear as yet.

In addition to its critical role in Notch signaling, ADAM 10 cleaves other biologically relevant substrates as well. Specifically, shedding of TNF-α (overlapping function with ADAM

17) and L1 [49] adhesion molecules as well as an in vitro type IV kollagenase [61] activity have been described, but the in vivo relevance of these interactions still awaits clarification.

In contrast, strong in vivo evidence was published for a critical role in axon extension,[62] the molecular nature of which is not fully clear as yet. Intriguing candidates would be ephrins (binding and cleavage of ephrinA2 is documented, Hattori et al 2000) and especially slit proteins and/or their receptors of the roundabout family.[63] With both options, a decisive role of ADAM 10 also in axon tract crossing could be envisaged also in vertebrates.

ADAM 17

The capacity of ADAM 17 for ectodomain shedding was first identified for the pro-inflammatory cytokine TNF-α[10] and the EGF-R ligand TGF-α.[12] The 'substrate list' was rapidly expanded in the following years, now including neuregulin,[64] but also amphiregulin as further EGF-R ligand. ADAM 17 also sheds EGF-R and the related erbB4 receptors themselves. Further established substrates for ADAM 17-mediated shedding are L-selectin,[65] TRANCE[66] and the chemokine fractalkine.[67]

For some of these substrates, a considerable overlap may exist with other ADAMs. A key example is the current controversy regarding the involvement of ADAMs 10 and 17 in the 'S2' cleavage of Notch receptors upon ligand binding (see above).

Previous data from invertebrate models clearly indicate a critical role of their respective ADAM 10 orthologues for this function.[42,58-60] Subsequent in vitro studies on murine cells were however unable to detect alterations in Notch processing in ADAM 10-deficient cells, but did so in ADAM 17 knockout fibroblasts.[3,5] In line with this unexpected result, the authors provided evidence for defective differentiation of ADAM 17-deficient bone marrow-derived monocytic progenitor cells. However, the overall phenotype of ADAM 17 knockout mice (see below) is unrelated to any of those obtained in models with a defective Notch signalling pathway. One explanation could be that due to differential tissue distribution of both ADAMs, a critical role for ADAM 17 in vivo is restricted to only a small number of cell types of low significance for overall development and integrity of an organism. It is thus highly likely that the current insight into Notch processing is subject to considerable change in the near future.

In line with such apparent overlaps in substrate specificity, the large number of potentially important substrates that can be cleaved by ADAM 17 is in striking contrast to the number of shedding events, for which ADAM 17 appears to be essential. This is best seen in ADAM 17 deficient mice.[12] It should be emphasized that in this mouse model specifically the zinc-binding site was targeted, inactivating the catalytic protease site of ADAM 17, but in all probability leaving the disintegrin and cystein-rich domains relevant for adhesive functions intact. These animals die between late prenatal development and the end of the first postnatal week, featuring a defective development of skin and lungs,[43] that can mostly be explained by the loss of TGF-α secretion (or, more broadly speaking, by a loss of soluble EGF-R ligands). The latter notion is supported by the observation that at least the defective lung development can in vitro be rescued by supplementing soluble EGF. The authors also report the loss or severe reduction of secretion of TNF receptor, L-selectin and TNF-α in their cell culture model. The contribution of these to the knockout phenotype still awaits further clarification.

Beyond the proteolytic functions discussed above and an as yet uncharacterized possible role of ADAM 17 as an adhesion molecule, interactions of its cytoplasmic domain with MAD2 have been described,[56] creating a link to cell cycle control of an as yet unknown significance.

With respect to APP secretion, ADAM 17 deficient fibroblasts have lost the capacity for phorbol ester-induced cleavage ("regulated" APPsα-secretion), whereas the constitutive secretion is not impaired. It should be noted that this stands in some contrast to results from transfection experiments cited above done with ADAMs 9 and 10, which indicated their involvement in both fractions of α-secretase activity. Also, studies on tissue localization of the different ADAMs consistently show that ADAM 17 is absent in most central neurons, further restricting its potential importance as APP α-secretase in the context of AD pathogenesis.

β-Secretase

The APP β-secretase has recently been identified by several groups in parallel as a new aspartyl protease, termed alternatively BACE 1, Asp 2 or memapsin 2.[27,29,32,68,69] A related enzyme, BACE 2, in addition exhibits an α-secretase like activity towards APP,[33,70,71] but is only weakly expressed in brain.

Almost in parallel with the identification of BACE 1, knockout mouse models[28,72] provided the good news that its loss abolished Aβ secretion in mice that appeared otherwise perfectly healthy. As the inhibition of γ-secretase, the other enzyme activity directly responsible for Aβ generation, is prone to cause severe side effects because it is also involved in the RIPping of several other decisive membrane proteins (see below), BACE 1 seems to be a highly attractive drug target.

Possibly in line with the lack of symptoms of BACE 1 knockout mice, the search for additional substrates of this protease so far did not turn up much. As yet, only the shedding of a Golgi resident sialyl transferase (ST6Gal) was shown in vitro to depend on BACE 1 activity.[73] As yet, the functional significance of ST6Gal secretion is unclear, but given the description of asymptomatic knockout mice, the role of BACE 1 is either compensated in vivo or may be functionally irrelevant.

However, the role of BACE 1 as an apparently ideal drug target to stall AD progression in a safe and effective manner is likely to cause an intense re-evaluation of BACE 1 deficient mice for more subtle pathology that might have been overlooked in the first investigations but could be severe enough to limit the use of BACE inhibitors in man. It is also an unanswered question, whether BACE 1 and BACE 2 could share a critical 'non-APP' substrate, as would be revealed by a double knockout model.

γ-Secretase

In contrast to α- and β-secretase activities, which are linked to single identifiable proteases, γ-secretase activity appears to be exerted by a multimeric protein complex that in addition to PSs at least also contains aph-2 / Nicastrin.[74] In this complex, PSs stand out by their alleged role as the proteolytically active center.[75] Their loss of function completely abolishes γ-secretase activity.[76]

γ-secretase functions as an i-Clip towards a rapidly growing number of substrates, which appear to share a type 1 topology as a common denominator.[9] Identified examples are the Notch receptor family, but also CD 44, erbB4,[77] the APLPs and possibly ATF 6 (but see refs.78) . Like in the case of the sheddases, the link between at least some of these substrates and PSs seems to be highly conserved in evolution, as demonstrated by the 'Notch phenotypes' in vertebrate and invertebrate PS deficient animals.[44,80,81]

Independent of their i-Clip function, PSs interact with other proteins. The best-characterized example is binding of β-catenin via the cytoplasmic loop domain, by which PSs downregulate its cytoplasmic concentration and thus the Wnt-mediated nuclear signaling. PS 1 is also a component of adherent junction complexes of synapses and endothelia[82] and cleaves E-cadherin in a γ-secretase—like fashion,[83] thereby dissociating membrane-associated β-catenin from the cytoskeleton and increasing the cytoplasmic concentration involved in Wnt signaling. Recently, telencephalin, a member of the ICAM family, has been added to the list of proteins being bound by PSs.[79]

When evaluated in a knockout mouse model, PS 1 and PS 2 appear to quite asymmetrically contribute to overall PS function, the vast majority of the γ-secretase activity depending on PS 1.[35] The developmental phenotype appears to be strikingly linked to impaired Notch signaling, which in the case of PS 1 deficient mice causes late prenatal death with cerebral hemorrhages, defective caudal somitogenesis,[84,85] cerebral and abdominal herniations and neuronal migration defects due to a premature death of Cajal-Retzius pioneer neurons.[86] This pattern is severely aggravated in PS1/PS2 double deficient mice that have lost any capacity for NICD generation,[37] as presenilins liberate the NICD from all Notch subtypes.[87] The embryos already die at E 9.5[38,88] somitogenesis now being abrogated after the first 4 to 5 rostralmost pairs, and yolk sac vasculature failing to reorganize in a pattern reminiscent of the situation in notch1/notch4 double knockout mice. Of note, PS 1 hypomorphic mice[89] likewise exhibited the axial patterning defects, but no brain abnormalities.

In contrast, PS 2 single deficient mice survive up to normal age,[88,90] but develop transient lung hemorrhages in early adulthood that become organized into a moderate lung fibrosis, identifying at least one organ in which PS 2 function cannot be replaced by PS 1.[38]

So far, the developmental PS knockout phenotype is interpreted as a 'Notch phenotype', fulfilling the pattern of defective vasculogenesis and segmentation also seen in a number of human Notch pathway mutants. It is as yet at best unclear, to what degree (if at all) an impairment of the above-mentioned adhesion function in endothelia could contribute to the hemorrhages. A specific contribution of other identified PS substrates and interacting proteins to a PS deficient phenotype has been characterized in PS1 knockout mice rescued with a PS1-encoding construct under the control of a Thy-1 promoter.[91] This promoter drives PS 1 expression in all major tissues except skin keratinocytes. In line with data from human disease, hyperactivity of this pathway has a tumor-promoting effect, causing a spectrum of malignant skin alterations in mice. However, it should be pointed out that these lesions develop only in adult mice, whereas for instance in 'unrescued' PS1 knockout mouse embryos the development of "catenin-type" malignancies like liver and colon carcinoma was not recorded so far. This indicates that besides the disturbed turnover of β-catenin, additional pathogenic events adding up over a considerable fraction of the murine life span are also required to elicit skin tumorigenesis.

Finally, PS function may be linked to 'non-Alzheimer' types of mental retardation. A recently described mutation of PS 1 (L113P,)[92] exhibiting a dominant–negative effect on γ-secretase activity both towards Notch and APP was linked to fronto-temporal dementia (FTD) with severe cortical atrophy. Clinical symptoms started around the age of 56 and severely deteriorated thereafter. The patient's family was analysed over four generations, and the PS 1 mutation fully cosegregated with the 6 clinically overt cases of FTD in this kindred. Most notably, a mutation of MAP τ (the only identified genetic cause of FTD so far) could not be detected. As the patient is still alive, histopathological data are not available. However, the secretion properties of the mutated PS1 make involvement of Aβ accumulations highly unlikely and have led the authors to the assumption of neurodegenerative effects caused e.g., by impaired Notch signaling to explain the frontal lobe atrophy.

Summary (Fig. 5.1)

Cleavage of APP is initiated by ectodomain shedding performed by either α- or β-secretase, creating the respective secreted fragments (APPsα or APPsβ) and transmembrane "stubs" of different lengths (C83 and C99). The stubs are further cleaved by γ-secretase releasing the APP intracellular domain (AICD) and the p3 or Aβ fragments, respectively.

The APP α-cleavage appears to be performed by several ADAM-type metalloproteases in parallel. All ADAMs in question are at least able to cleave a number of other substrates as well and may also have cell adhesion function. However, knockout data indicate that for most of the identified substrate of an ADAM protease other sheddases exist. For instance, the phenotype of the ADAM 17 deficient mouse is characterized by the loss of TGF-α secretion.

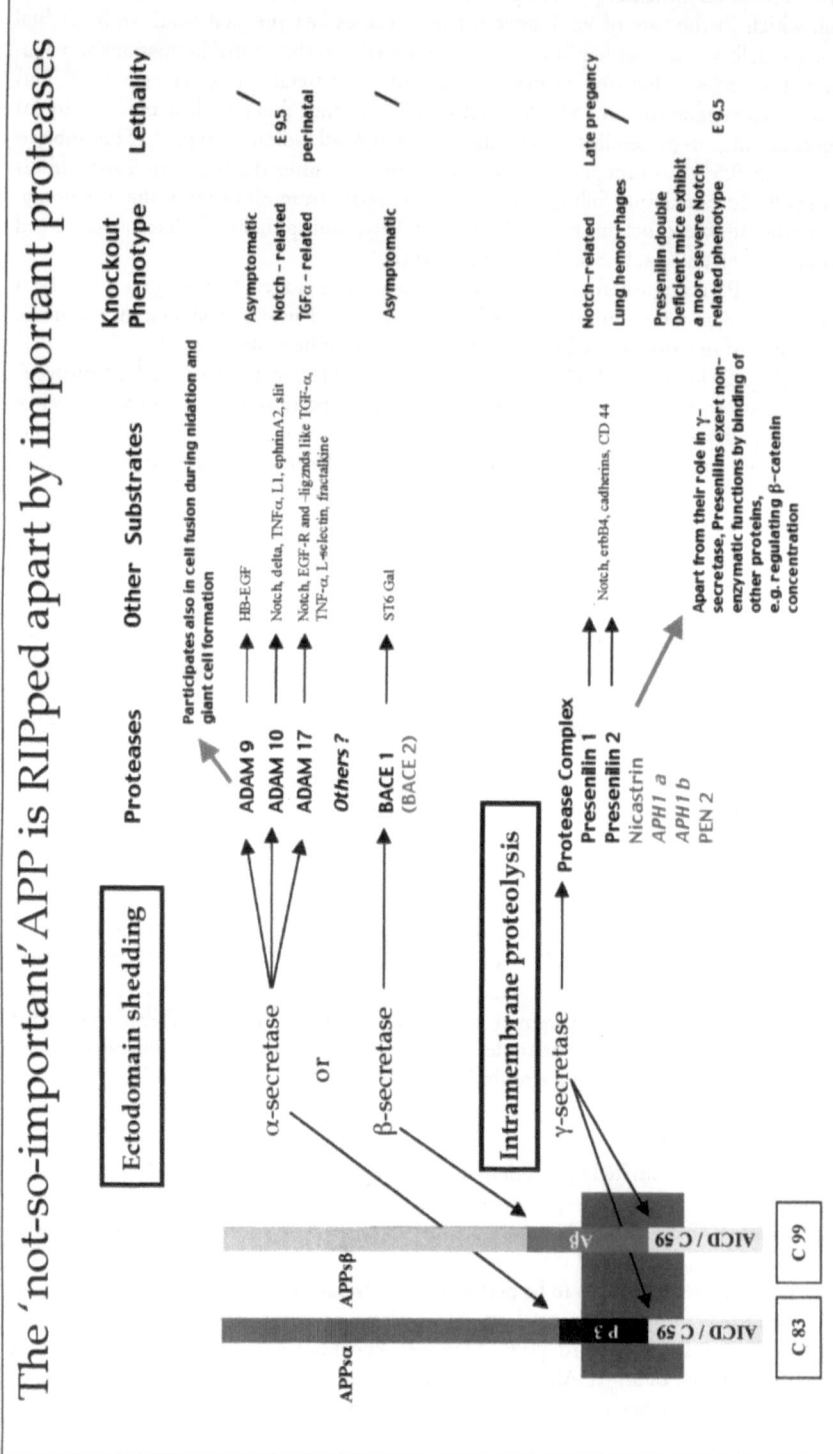

Figure 5.1. Schematic representation of (candidate) APP secretases and their role in animal development..

For BACE1, only a sialyl transferase has been identified as an additional shed substrate; however, the functional significance is unclear. Within the multiprotein complex of γ-secretase, protease activity is exerted in all probability by the presenilins. Their loss of function causes a phenotype mainly explained by an incapacitated Notch pathway. However, presenilins have additional nonenzymatic functions, for instance in the regulation of the Wnt/β-catenin pathway. Their impairment consecutive to a partially rescued presenilin knockout causes symptoms only during adulthood, but apparently does not contribute to developmental pathology.

References

1. De Strooper B, Annaert W. Proteolytic processing and cell biological functions of the amyloid precursor protein. J Cell Sci 2000; 113:1857-70.
2. Selkoe DJ. Alzheimer's disease: genes, proteins, and therapy. Physiol Rev 2001; 81:741-66.
3. Brou C, Logeat F, Gupta N et al. A novel proteolytic cleavage involved in Notch signaling: the role of the disintegrin-metalloprotease TACE. Mol Cell 2000; 5:207-16.
4. Hartmann D, Tournoy J, Saftig P et al. Implication of APP secretases in notch signaling. J Mol Neurosci 2001; 17:171-81.
5. Mumm JS, Schroeter EH, Saxena MT et al. A ligand-induced extracellular cleavage regulates gamma-secretase-like proteolytic activation of Notch1. Mol Cell 2000; 5:197-206.
6. Nohturfft A, DeBose-Boyd RA, Scheek S et al. Sterols regulate cycling of SREBP cleavage-activating protein (SCAP) between endoplasmic reticulum and Golgi. Proc Natl Acad Sci USA 1999; 96:11235-40.
7. Nohturfft A, Yabe D, Goldstein JL et al. Regulated step in cholesterol feedback localized to budding of SCAP from ER membranes. Cell 2000; 102:315-23.
8. Niwa M, Sidrauski C, Kaufman RJ, Walter P. A role for presenilin-1 in nuclear accumulation of Ire1 fragments and induction of the mammalian unfolded protein response [In Process Citation]. Cell 1999; 99:691-702.
9. Brown MS, Ye J, Rawson RB, Goldstein JL. Regulated intramembrane proteolysis: a control mechanism conserved from bacteria to humans. Cell 2000; 100:391-8.
10. Black RA, Rauch CT, Kozlosky CJ et al. A metalloproteinase disintegrin that releases tumour-necrosis factor-alpha from cells. Nature 1997; 385:729-33.
11. Hooper NM, Karran EH, Turner AJ. Membrane protein secretases. Biochem J 1997; 321(Pt.2):265-79.
12. Peschon JJ, Slack JL, Reddy P et al. An essential role for ectodomain shedding in mammalian development. Science 1998; 282:1281-4.
13. Schlondorff J, Blobel CP. Metalloprotease-disintegrins: modular proteins capable of promoting cell-cell interactions and triggering signals by protein-ectodomain shedding. J Cell Sci 1999; 112:3603-17.
14. Heber S, Herms J, Gajic V et al. Mice with combined gene knock-outs reveal essential and partially redundant functions of amyloid precursor protein family members. J Neurosci 2000; 20:7951-63.
15. von Koch CS, Zheng H, Chen H et al. Generation of APLP2 KO mice and early postnatal lethality in APLP2/APP double KO mice. Neurobiol Aging 1997; 18:661-9.
16. Cupers P, Orlans I, Craessaerts K et al. The amyloid precursor protein (APP)-cytoplasmic fragment generated by gamma-secretase is rapidly degraded but distributes partially in a nuclear fraction of neurones in culture. J Neurochem 2001; 78:1168-78.
17. Cao X, Sudhof TC. A transcriptionally [correction of transcriptively] active complex of APP with Fe65 and histone acetyltransferase Tip60. Science 2001; 293:115-20.
18. Sabo SL, Ikin AF, Buxbaum JD, Greengard P. The Alzheimer amyloid precursor protein (APP) and FE65, an APP-binding protein, regulate cell movement. J Cell Biol 2001; 153:1403-14.
19. Zambrano N, Bimonte M, Arbucci S et al. feh-1 and apl-1, the Caenorhabditis elegans orthologues of mammalian Fe65 and beta-amyloid precursor protein genes, are involved in the same pathway that controls nematode pharyngeal pumping. J Cell Sci 2002; 115:1411-22.
20. Zambrano N, Bruni P, Minopoli G et al. The beta-amyloid precursor protein APP is tyrosine-phosphorylated in cells expressing a constitutively active form of the Abl protoncogene. J Biol Chem 2001; 276:19787-92.

21. Esler WP, Wolfe MS. A portrait of Alzheimer secretases--new features and familiar faces. Science 2001; 293:1449-54.

22. Koike H, Tomioka S, Sorimachi H et al. Membrane-anchored metalloprotease MDC9 has an alpha-secretase activity responsible for processing the amyloid precursor protein. Biochem J 1999; 343(Pt.2):371-5.

23. Kojro E, Gimpl G, Lammich S et al. Low cholesterol stimulates the nonamyloidogenic pathway by its effect on the alpha -secretase ADAM 10. Proc Natl Acad Sci USA 2001; 98:5815-20.

24. Lammich S, Kojro E, Postina R et al. Constitutive and regulated alpha-secretase cleavage of Alzheimer's amyloid precursor protein by a disintegrin metalloprotease [In Process Citation]. Proc Natl Acad Sci USA 1999; 96:3922-7.

25. Fahrenholz F, Gilbert S, Kojro E et al. Alpha-secretase activity of the disintegrin metalloprotease ADAM 10. Influences of domain structure. Ann NY Acad Sci 2000; 920:215-22.

26. Buxbaum JD, Liu KN, Luo Y et al. Evidence that tumor necrosis factor alpha converting enzyme is involved in regulated alpha-secretase cleavage of the Alzheimer amyloid protein precursor. J Biol Chem 1998; 273:27765-7.

27. Cai H, Wang Y, McCarthy D et al. BACE1 is the major beta-secretase for generation of Abeta peptides by neurons. Nat Neurosci 2001; 4:233-4.

28. Luo Y, Bolon B, Kahn S et al. Mice deficient in BACE1, the Alzheimer's beta-secretase, have normal phenotype and abolished beta-amyloid generation. Nat Neurosci 2001; 4:231-2.

29. Vassar R, Bennett BD, Babu-Khan S et al. Beta-secretase cleavage of Alzheimer's amyloid precursor protein by the transmembrane aspartic protease BACE. Science 1999; 286:735-41.

30. Vassar R, Citron M. Abeta-generating enzymes: recent advances in beta- and gamma-secretase research. Neuron 2000; 27:419-22.

31. Farzan M, Schnitzler CE, Vasilieva N et al. BACE2, a beta -secretase homolog, cleaves at the beta site and within the amyloid-beta region of the amyloid-beta precursor protein. Proc Natl Acad Sci USA 2000; 97:9712-7.

32. Hussain I, Powell DJ, Howlett DR et al. ASP1 (BACE2) cleaves the amyloid precursor protein at the beta-secretase site. Mol Cell Neurosci 2000; 16:609-19.

33. Solans A, Estivill X, de LLS. A new aspartyl protease on 21q22.3, BACE2, is highly similar to Alzheimer's amyloid precursor protein beta-secretase. Cytogenet Cell Genet 2000; 89:177-84.

34. Yan R, Munzner JB, Shuck ME, Bienkowski MJ. BACE2 functions as an alternative alpha-secretase in cells. J Biol Chem 2001; 276:34019-27.

35. De Strooper B, Saftig P, Craessaerts K et al. Deficiency of presenilin-1 inhibits the normal cleavage of amyloid precursor protein. Nature 1998; 391:387-90.

36. Haass C, De SB. The presenilins in Alzheimer's disease--proteolysis holds the key. Science 1999; 286:916-9.

37. Herreman A, Serneels L, Annaert W et al. Total inactivation of gamma-secretase activity in presenilin-deficient embryonic stem cells. Nat Cell Biol 2000; 2:461-2.

38. Herreman A, Hartmann D, Annaert W et al. Presenilin 2 deficiency causes a mild pulmonary phenotype and no changes in amyloid precursor protein processing but enhances the embryonic lethal phenotype of presenilin 1 deficiency. Proc Natl Acad Sci USA 1999; 96:11872-7.

39. Kimberly WT, Xia W, Rahmati T et al. The transmembrane aspartates in presenilin 1 and 2 are obligatory for gamma-secretase activity and amyloid beta-protein generation. J Biol Chem 2000; 275:3173-8.

40. Li YM, Lai MT, Xu M et al. Presenilin 1 is linked with gamma-secretase activity in the detergent solubilized state. Proc Natl Acad Sci USA 2000; 97:6138-43.

41. Rooke J, Pan D, Xu T, Rubin GM. KUZ, a conserved metalloprotease-disintegrin protein with two roles in Drosophila neurogenesis. Science 1996; 273:1227-31.

42. Wen C, Metzstein MM, Greenwald I. SUP-17, a Caenorhabditis elegans ADAM protein related to Drosophila KUZBANIAN, and its role in LIN-12/NOTCH signalling. Development 1997; 124:4759-67.

43. Zhao J, Chen H, Peschon JJ et al. Pulmonary hypoplasia in mice lacking tumor necrosis factor-alpha converting enzyme indicates an indispensable role for cell surface protein shedding during embryonic lung branching morphogenesis. Dev Biol 2001; 232:204-18.

44. Ye Y, Lukinova N, Fortini ME. Neurogenic phenotypes and altered Notch processing in Drosophila Presenilin mutants. Nature 1999; 398:525-9.

45. De Strooper B, Annaert W, Cupers P et al. A presenilin-1-dependent gamma-secretase-like protease mediates release of Notch intracellular domain. Nature 1999; 398:518-22.
46. Wong PC, Zheng H, Chen H et al. Presenilin 1 is required for Notch1 and DII1 expression in the paraxial mesoderm. Nature 1997; 387:288-92.
47. Evans JP. Fertilin beta and other ADAMs as integrin ligands: insights into cell adhesion and fertilization. Bioessays 2001; 23:628-39.
48. Huovila AP, Almeida EA, White JM. ADAMs and cell fusion. Curr Opin Cell Biol 1996; 8:692-9.
49. Mechtersheimer S, Gutwein P, Agmon-Levin N et al. Ectodomain shedding of L1 adhesion molecule promotes cell migration by autocrine binding to integrins. J Cell Biol 2001; 155:661-73.
50. Alfandari D, Cousin H, Gaultier A et al. Xenopus ADAM 13 is a metalloprotease required for cranial neural crest-cell migration. Curr Biol 2001; 11:918-30.
51. Lee HJ, Jung KM, Huang YZ et al. Presenilin-dependent {gamma}-secretase-like intramembrane cleavage of ErbB4. J Biol Chem 2001.
52. Weskamp G, Cai H, Brodie TA et al. Mice lacking the metalloprotease-disintegrin MDC9 (ADAM9) have no evident major abnormalities during development or adult life. Mol Cell Biol 2002; 22:1537-44.
53. Inoue D, Reid M, Lum L et al. Cloning and initial characterization of mouse meltrin beta and analysis of the expression of four metalloprotease-disintegrins in bone cells. J Biol Chem 1998; 273:4180-7.
54. Mahimkar RM, Baricos WH, Visaya O et al. Identification, cellular distribution and potential function of the metalloprotease-disintegrin MDC9 in the kidney. J Am Soc Nephrol 2000; 11:595-603.
55. Olson GE, Winfrey VP, Matrisian PE et al. Blastocyst-dependent upregulation of metalloproteinase/disintegrin MDC9 expression in rabbit endometrium. Cell Tissue Res 1998; 293:489-98.
56. Nelson KK, Schlondorff J, Blobel CP. Evidence for an interaction of the metalloprotease-disintegrin tumour necrosis factor alpha convertase (TACE) with mitotic arrest deficient 2 (MAD2), and of the metalloprotease-disintegrin MDC9 with a novel MAD2-related protein, MAD2beta. Biochem J 1999; 343 Pt 3:673-80.
57. Asakura M, Kitakaze M, Takashima S et al. Cardiac hypertrophy is inhibited by antagonism of ADAM12 processing of HB-EGF: metalloproteinase inhibitors as a new therapy. Nat Med 2002; 8:35-40.
58. Pan D, Rubin GM. Kuzbanian controls proteolytic processing of Notch and mediates lateral inhibition during Drosophila and vertebrate neurogenesis. Cell 1997; 90:271-80.
59. Sotillos S, Roch F, Campuzano S. The metalloprotease-disintegrin Kuzbanian participates in Notch activation during growth and patterning of Drosophila imaginal discs. Development 1997; 124:4769-79.
60. Lieber T, Kidd S, Young MW. kuzbanian-mediated cleavage of Drosophila Notch. Genes Dev 2002; 16:209-21.
61. Millichip MI, Dallas DJ, Wu E et al. The metallo-disintegrin ADAM10 (MADM) from bovine kidney has type IV collagenase activity in vitro. Biochem Biophys Res Commun 1998; 245:594-8.
62. Fambrough D, Pan D, Rubin GM, Goodman CS. The cell surface metalloprotease/disintegrin Kuzbanian is required for axonal extension in Drosophila. Proc Natl Acad Sci USA 1996; 93:13233-8.
63. Schimmelpfeng K, Gogel S, Klambt C. The function of leak and kuzbanian during growth cone and cell migration. Mech Dev 2001; 106:25-36.
64. Montero JC, Yuste L, Diaz-Rodriguez E et al. Differential shedding of transmembrane neuregulin isoforms by the tumor necrosis factor-alpha-converting enzyme. Mol Cell Neurosci 2000; 16:631-48.
65. Condon TP, Flournoy S, Sawyer GJ et al. ADAM17 but not ADAM10 mediates tumor necrosis factor-alpha and L-selectin shedding from leukocyte membranes. Antisense Nucleic Acid Drug Dev 2001; 11:107-16.
66. Lum L, Wong BR, Josien R et al. Evidence for a role of a tumor necrosis factor-alpha (TNF-alpha)-converting enzyme-like protease in shedding of TRANCE, a TNF family member involved in osteoclastogenesis and dendritic cell survival. J Biol Chem 1999; 274:13613-8.
67. Garton KJ, Gough PJ, Blobel CP et al. Tumor necrosis factor-alpha-converting enzyme (ADAM17) mediates the cleavage and shedding of fractalkine (CX3CL1). J Biol Chem 2001; 276:37993-8001.

68. Lin X, Koelsch G, Wu S et al. Human aspartic protease memapsin 2 cleaves the beta-secretase site of beta-amyloid precursor protein. Proc Natl Acad Sci USA 2000; 97:1456-60.

69. Yan R, Bienkowski MJ, Shuck ME et al. Membrane-anchored aspartyl protease with Alzheimer's disease beta-secretase activity. Nature 1999; 402:533-7.

70. Bennett BD, Babu-Khan S, Loeloff R et al. Expression analysis of BACE2 in brain and peripheral tissues. J Biol Chem 2000; 275:20647-51.

71. Acquati F, Accarino M, Nucci C et al. The gene encoding DRAP (BACE2), a glycosylated transmembrane protein of the aspartic protease family, maps to the down critical region. FEBS Lett 2000; 468:59-64.

72. Roberds SL, Anderson J, Basi G et al. BACE knockout mice are healthy despite lacking the primary beta-secretase activity in brain: implications for Alzheimer's disease therapeutics. Hum Mol Genet 2001; 10:1317-24.

73. Kitazume S, Tachida Y, Oka R et al. Alzheimer's beta-secretase, beta-site amyloid precursor protein-cleaving enzyme, is responsible for cleavage secretion of a Golgi-resident sialyltransferase. Proc Natl Acad Sci USA 2001; 98:13554-9.

74. Hu Y, Ye Y, Fortini ME. Nicastrin Is Required for gamma-Secretase Cleavage of the Drosophila Notch receptor. Dev Cell 2002; 2:69-78.

75. Wolfe MS, Haass C. The Role of presenilins in gamma-secretase activity. J Biol Chem 2001; 276:5413-6.

76. Herreman A, Serneels L, Annaert W et al. Total inactivation of gamma-secretase activity in presenilin-deficient embryonic stem cells. Nat Cell Biol 2000; 2:461-2.

77. Lee HJ, Jung KM, Huang YZ et al. Presenilin-dependent gamma-secretase-like intramembrane cleavage of ErbB4. J Biol Chem 2002; 277:6318-23.

78. Steiner H, Winkler E, Shearman MS et al. Endoproteolysis of the ER stress transducer ATF6 in the presence of functionally inactive presenilins. Neurobiol Dis 2001; 8:717-22.

79. Annaert WG, Esselens C, Baert V et al. Interaction with telencephalin and the amyloid precursor protein predicts a ring structure for presenilins. Neuron 2001; 32:579-89.

80. Li X, Greenwald I. HOP-1, a caenorhabditis elegans presenilin, appears to be functionally redundant with SEL-12 presenilin and to facilitate LIN-12 and GLP-1 signaling [In Process Citation]. Proc Natl Acad Sci USA 1997; 94:12204-9.

81. Levitan D, Greenwald I. Effects of SEL-12 presenilin on LIN-12 localization and function in caenorhabditis elegans [In Process Citation]. Development 1998; 125:3599-606.

82. Georgakopoulos A, Marambaud P, Friedrich VL, Jr. et al. Presenilin-1: a component of synaptic and endothelial adherens junctions. Ann NY Acad Sci 2000; 920:209-14.

83. Marambaud P, Shioi J, Serban G et al. A presenilin-1/gamma-secretase cleavage releases the E-cadherin intracellular domain and regulates disassembly of adherens junctions. Embo J 2002; 21:1948-56.

84. Shen J, Bronson RT, Chen DF et al. Skeletal and CNS defects in Presenilin-1-deficient mice. Cell 1997; 89:629-39.

85. Wong PC, Zheng H, Chen H et al. Presenilin 1 is required for Notch1 and DII1 expression in the paraxial mesoderm. Nature 1997; 387:288-92.

86. Hartmann D, Strooper BD, Saftig P. Presenilin-1 deficiency leads to loss of Cajal-Retzius neurons and cortical dysplasia similar to human type 2 lissencephaly. Curr Biol 1999; 9:719-27.

87. Saxena MT, Schroeter EH, Mumm JS, Kopan R. Murine notch homologs (N1-4) undergo presenilin-dependent proteolysis. J Biol Chem 2001; 276:40268-73.

88. Donoviel DB, Hadjantonakis AK, Ikeda M et al. Mice lacking both presenilin genes exhibit early embryonic patterning defects. Genes Dev 1999; 13:2801-10.

89. Rozmahel R, Huang J, Chen F et al. Normal brain development in PS1 hypomorphic mice with markedly reduced gamma-secretase cleavage of betaAPP. Neurobiol Aging 2002; 23:187-94.

90. Steiner H, Duff K, Capell A et al. A loss of function mutation of presenilin-2 interferes with amyloid beta-peptide production and notch signaling. J Biol Chem 1999; 274:28669-73.

91. Xia X, Qian S, Soriano S et al. Loss of presenilin 1 is associated with enhanced beta-catenin signaling and skin tumorigenesis. Proc Natl Acad Sci USA 2001; 98:10863-8.

92. Amtul Z, Lewis PA, Piper S et al. A presenilin 1 mutation associated with familial frontotemporal dementia inhibits gamma-secretase cleavage of APP and notch. Neurobiol Dis 2002; 9:269-73.

CHAPTER 6

Proteolytic Degradation of Aβ by Neprilysin and Other Peptidases

Takaomi C. Saido and Hiroyuki Nakahara

Abstract

Amyloid β peptide (Aβ), the pathogenic agent of Alzheimer's disease (AD), is a physiological metabolite in the brain. We focused our attention and effort on elucidation of the unresolved aspect of Aβ metabolism, proteolytic degradation. Among a number of Aβ-degrading enzyme candidates, we identified a member of the neutral endopeptidase family, neprilysin, as the major catabolic enzyme using a novel in vivo paradigm. Consistently, neprilysin deficiency resulted in defects in the metabolism of endogenous Aβ 40 and 42, notably, in a gene dose-dependent manner. Our observations suggest that even partial down-regulation of neprilysin activity, which could be caused by aging, can contribute to AD development by promoting Aβ accumulation. Moreover, we describe that an aging-dependent decline of neprilysin activity, leading to elevation of Aβ levels in the brain is a natural process preceding AD pathology. We not only hypothesize in this chapter on the likely role of neprilysin down-regulation in sporadic AD (SAD) pathogenesis but also propose that the knowledge itself be used for developing novel preventive and therapeutic strategies. Finally, we also discuss the need for mathematical refinement of the current human genetic approaches to identify the risk and anti-risk factors for AD.

Introduction: Why Aβ Degradation?

As discussed in Chapter 1, the steady-state Aβ levels are determined by the dynamic and steady-state metabolic balances between the generation and clearance. The clearance mechanism consists of proteolytic degradation within brain parenchyma and transport to the cerebrospinal fluid (CSF) and to plasma followed by degradation outside the brain. In our laboratory, we chose to focus our attention and efforts on elucidation of the in-parenchyma degradation[1,2] (see Fig. 6.1), assuming that the rate constants in Formula 1.1 (Chapter 1) were such that $K_2 > K_3$ and thus $[A\beta_{42}] \approx K_1/K_2 \times [APP]$ (almost). The basis of our reasoning is as follows:

1. Brains are rich sources of various peptidases which can proteolyze Aβ as long as they are accessible to the substrate.
2. Because the amounts of Aβ in the brain and in the CSF or plasma are poorly correlated with each other in AD patients,[3,4] there does not seem to exist an effective transport-dependent mechanism that clears Aβ out of the brain at least in humans under the pathological conditions.

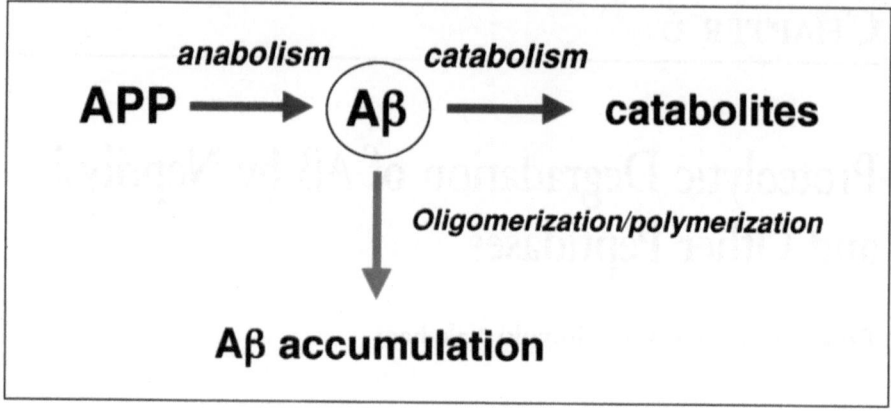

anabolism   catabolism

APP ⟶ **Aβ** ⟶ **catabolites**

Oligomerization/polymerization

Aβ accumulation

Figure 6.1. Relationships between Aβ anabolism, catabolism, and pathological accumulation.
Aβ is a physiological peptide constantly anabolized and immediately catabolized before being deposited in the brain under normal conditions. The clearance mechanism involving the transport out of brain to the circulatory system (see Fig. 1.5) is excluded in this scheme.

3. If reduced efficiency of transport out of brain to the circulatory system was a major cause of the pathological Aβ deposition, one would expect vascular amyloid deposition to arise earlier than parenchymal deposition, but the reality is generally the other way around particularly in humans.
4. Human brains are approximately 1,000 times larger than mouse brains in size. Therefore, the advantage of in-parenchyma degradation over the transport-dependent clearance is likely to be even greater in humans than in mice particularly if the major transport pathway is via the interstitial fluid (ISF) and then to the CSF and plasma. Note that the primary function of proteolysis is recycling of amino acids, therefore in-parenchyma degradation is more economical from an energy metabolic point of view than out-of-brain transport especially in the case of humans.
5. Because increases in Aβ production cause familial AD (FAD), leading to elevation of steady-state Aβ levels in the brain,[5] and because Aβ overproduction is rarely observed in sporadic AD (SAD) cases, decreases in Aβ degradation could be the cause for the majority of SAD cases.[4]

These reasons do not necessarily rule out the importance of the transport mechanism. This is because, as discussed by Zlokovic and Holtzman in later chapters (Chapters 10 and 11), a dynamic balance seems to exist between the bran and plasma through the transport across the blood-brain barrier (BBB) and/or CSF under certain conditions at least in rodent models (particularly, when high-affinity anti-Aβ antibodies are injected into the plasma).[6,7] If this also applies to humans to the extent that affects the risk of developing AD, physiological steady-state plasma and/or CSF Aβ levels could be an important parameter, and thus we may have to view the Aβ-degradation issue in a more systemic manner rather than focusing only on the processes within the brain. Besides, it is quite possible that the mechanism closely associated with CAA formation may involve the transport mechanism rather than the in-parenchyma degradation. The latter point, however, is beyond the scope of this chapter.

Two Major Categories of Aβ-Degrading Mechanisms

The mechanism of Aβ degradation can be classified into at least two categories: "physiological degradation" that determines normal steady-state levels and "pathological degradation," which may be mobilized to counter-balance the increased Aβ levels after the deposition starts. The distinction between these two categories is not necessarily clear, because peptidase(s) in the first category is likely to be involved in the second category degradation as well through alteration of the dynamic balances between monomers, oligomers and polymers of Aβ.

Although both categories are probably equally important in general terms, our impression is that the first category degradation is more likely to be associated with the etiology of SAD because the common in vivo phenotype of almost all the FAD-causing mutations is elevation of the steady-state Aβ$_{42}$ levels in brain.[5] We however had previously hypothesized that the plasmin system and matrix metalloproteinase (MMP) system might be candidates that could dispose of polymerized Aβ because they are the major proteinases capable of degrading fibrillized proteins.[8] Consequently, we identified a novel brain-specific MMP, MT-5 MMP.[9] Although our anti-MT5-MMP antibodies do in fact stain both senile plaques and neurofibrillary tangles in AD patients.[9] We noted that NFT is more strongly stained (Iwata and Saido, unpublished data) and this was confirmed by Marion Maat-Schieman, Leiden University Medical Center, in hereditary cerebral hemorrhage with amyloidosis Dutch-type (HCHWA-D) brains and also by Takeshi Iwatsubo, University of Tokyo, in SAD brains (personal communications). Because MT-5 is a type 1 membrane-spanning protein, we do not yet know how important this proteinase is in AD pathogenesis. This project is presently suspended in our laboratory mainly because we did not succeed in observing any proteolytic action of MT-5 MMP on fibrillized Aβ even in an in vitro paradigm, because both the plasmin and MMP systems generally require activation of latent precursors, and because proteinases belonging to these systems could be destructive by degrading a number of matrix-consisting proteins and could worsen the pathological states through, for instance, inducing inflammation.

We also need to consider such cellular degradation mechanism as phagocytosis and pinocytosis conducted primarily by activated astrocytes and microglia.[10,11] I left out this important cellular aspect of Aβ clearance just to avoid complexity. In Chapter 12, Cindy Lemere et al refer to this aspect of Aβ clearance in association with the vaccination topic. For further details, see reviews and articles published by such experts as P. McGeer, G. Cole, F. Maxwell, T. Weiss-Corary, etc.

Identification of Neprilysin as a Major in vivo Aβ-Degrading Enzyme in Brain

A number of reports have described potential Aβ-degrading enzymes using either test-tube or in vitro paradigms. See Table 6.1 for the list of major candidates thus far described.[12-33] The intrinsic problem of such paradigms is that, if you incubate Aβ with a given peptidase under the optimal conditions for long enough, almost any peptidase will degrade the substrate. Our quest to identify the physiologically relevant in vivo peptidase(s) was based on the philosophy expressed by the following metaphor.

"If you feed a hungry lion (almost any protease) in a cage (test tube or cell culture) a penguin (Aβ), the lion would probably eat the penguin, but such an observation obviously does not tell you that lions are natural predators of penguins. You have to capture the process in an environment that is as natural as possible (in the brain of live animals) without affecting the environment." This metaphor is not meant to be sarcastic to scientists who have been mainly employing in vitro paradigms. It is just that the degradation problems cannot be dealt with properly in the same way as the production problems have been, for which the molecular and cellular approaches have indeed been powerful and relevant tools. Besides, it is not exactly our original metaphor but rather is partly a form of cultural transmission. The late Efraim Racker used to tell TCS something like "If you have sufficient amounts of a kinase, ATP, and necessary cofactors, you can even phosphorylate your grandmother," when TCS was a visiting graduate student to Cornell University, working with a biophysicist, Watt W. Webb, in 1985-1986. Another factor that makes degradation research more difficult to perform is that the major

Table 6.1. List of some candidates for Aβ-degrading enzymes

Cathepsin D
Insulin-degrading enzyme (IDE)
Plasmin
Neprilysin
Endothelin-converting enzyme (ECE)
Angiotensin-converting enzyme (ACE)
Matrix metalloproteinase 2 (MMP-2)
Matrix metalloproteinase 9 (MMP-9)
Coagulation factor XIa
Carboxypeptidase
Metalloendopeptidase (unidentified)
Serinepeptiade (unidentified)

degradation processes are likely to take place extracytoplasmically, i.e., in extracellular space or on the lumen side of intracellular vesicles represented by secretory vesicles axonally transported from soma to presynaptic terminals. This means that not only the cellular topology but also the complex structural organization of the brain tissue composed of various types of cells need to be taken into consideration.

Therefore, we started our series of degradation studies by establishing a novel in vivo experimental paradigm in which we injected synthetic internally multi-radio-labeled $Aβ_{1-42}$ in to the hippocampus of anesthetized live rats and analyzed the degradation process by high-performance liquid chromatography (HPLC) directly connected a flow-type scintillation counter (Fig. 6.2).[1] Experiments using a panel of more than 20 peptidase inhibitors suggested that a neutral endopeptidase family member similar or identical to neprilysin appears to play a major role in the $Aβ_{1-42}$ catabolism because thiorphan, a well-characterized neutral endopeptidase inhibitor, was the most potent inhibitor. In accordance, short-term and long-term infusions of rat hippocampus with thiorphan resulted in biochemical and pathological accumulation of endogenous Aβ, respectively.

The next task was to identify the major responsible peptidase among members of the neutral endopeptidase family. We thus made efforts to determine its molecular identity by biochemical and molecular biological approaches and, consequently, predicted that neprilysin is likely to be the primary candidate.[34,35] We thus examined the degradation of the radio-labeled Aβ in neprilysin-KO mouse brains using their wild-type litter mates as positive controls and confirmed our prediction as follows.[2]

Neprilysin is a type II membrane-associated peptidase with the active site facing the lumen or extracellular side of membranes[36-39] (Fig. 6.3). This topology is favorable for the degradation of extracytoplasmic peptides such as Aβ. Furthermore, we devised to establish quantitative immunofluorescence visualization of neprilysin using neprilysin-KO[40] mice as a negative control. (We actually had to test several tens of combinations of antibodies and protocols to optimize the experimental protocols). We have confirmed that neprilysin is essentially exclusively expressed in neurons, not in glia, and that the peptidase, after synthesis in the soma, is axonally transported to presynaptic terminals[40] presumably in a manner similar to the way APP is transported. Therefore, presynaptic terminals and nearby intracellular (lumen-sided) locations are likely to be the sites of Aβ degradation by neprilysin (Fig. 6.4).

Based on the above observations, we examined the ability of neprilysin-KO mouse brains to degrade Aβ in vivo and quantified the endogenous Aβ levels in the brains.[2] Due presumably to the redundancy in the neutral endopeptidase family,[35,38,41] the KO mice are normal in

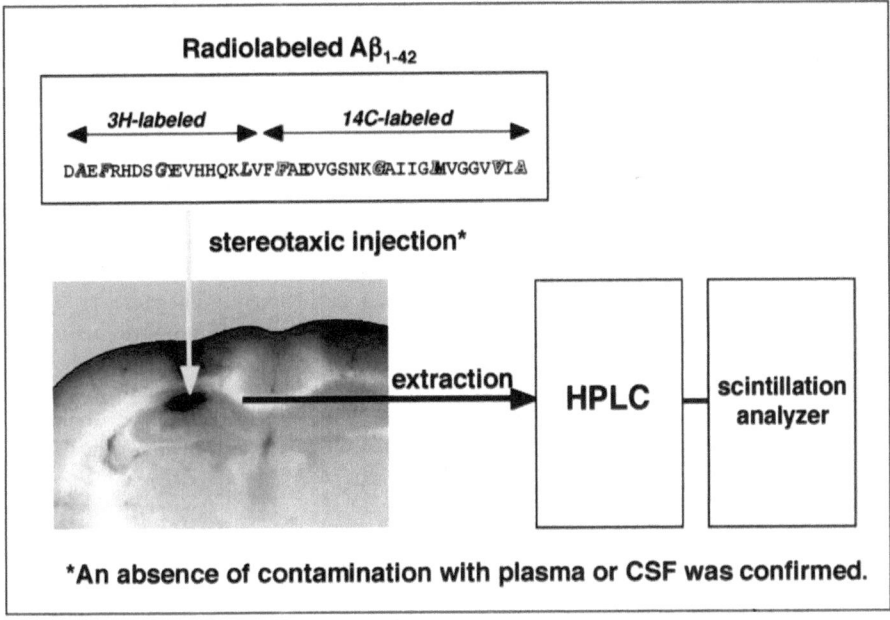

Figure 6.2. The initial experimental paradigm used for analyzing in vivo Aβ degradation in the hippocampus. We established a novel in vivo experimental paradigm in which we injected synthetic internally multi-radio-labeled Aβ$_{1-42}$ into the hippocampus of anesthetized live rats and analyzed the degradation process by high-performance liquid chromatography (HPLC) directly connected a flow-type scintillation counter. See the section on Identification of Neprilysin as a Major in vivo Aβ-Degrading Enzyme in Brain and ref. 1 for details.

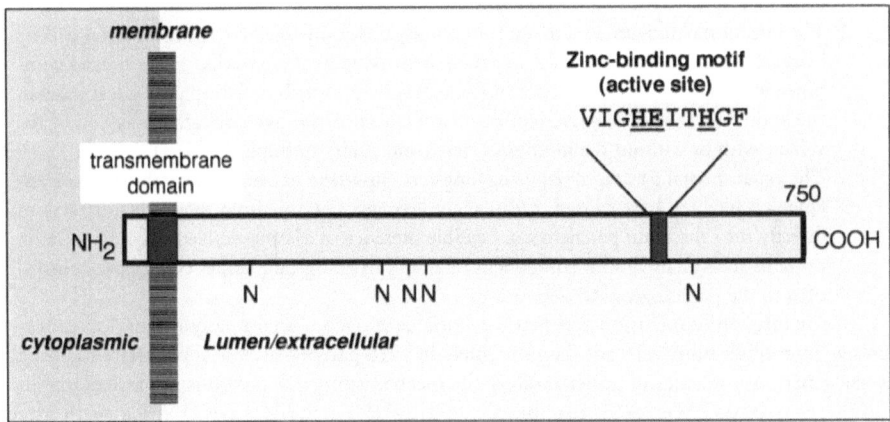

Figure 6.3. Schematic structure and cellular topology of neprilysin.
Neprilysin is a type II membrane-associated peptidase with the active site facing the lumen or extracellular side of membranes. The enzyme requires zinc for its activity. N: N-glycosylation sites. The glycosylation is reported also to be required for the enzymatic activity to be fully expressed.[70]

reproduction, development, and adult anatomy to our knowledge.[42] The ability to degrade the radiolabeled Aβ (see Fig. 6.2) was significantly reduced in the KO mouse brains. Consistently, both the endogenous $A\beta_{40}$ and $A\beta_{42}$ levels were elevated approximately 2-fold, in a manner comparable to or even greater than what has been described in FAD-causing mutant presenilin transgenic or knock-in mice[43-46] (Fig. 6.5). More importantly, the elevation of Aβ levels was inversely correlated with the gene dose of neprilysin and thus with the enzyme activity (Fig. 6.6). These observations suggest that even partial loss of neprilysin expression/activity causes the elevation of $A\beta_{42}$ levels and thus could induce Aβ amyloidosis on a long-term basis in a manner similar to that of FAD-causing gene mutations. These results also indicate that the rate constant for the in-parenchyma degradation conducted by neprilysin accounts for about 50% of all the clearance activity, or $K_2 + K_3$ in Formula 1.1 in Chapter 1. Therefore, K_2 being negligibly smaller than K_3 is unlikely and K_2 is likely to be even greater in human brains for fourth of the reasons stated in the Introduction section.

In a similar manner, some of the other Aβ-degrading enzyme candidates were examined by a similar reverse genetic approach, i.e., by measuring the brain Aβ levels in the brains of KO-mice (Table. 6.2). To our knowledge, neprilysin seems to be the only peptidase that actually regulates the steady-state level of the primarily pathogenic Aβ species, $A\beta_{42}$, as the pathogenic FAD presenilin mutations do. This however does not necessarily rule out these other peptidases as relevant Aβ degrading enzymes in more general terms (see section *Other Aβ-Degrading Enzyme Candidates* in this chapter and Chapter 7).

Although we previously described the pathological deposition of the endogenous Aβ in mouse brains infused with thiorphan,[1] we never observed such pathological structures in the brains of neprilysin-deficient mouse brains.[2] Several reasons for this apparent discrepancy can be advanced as follows.

1. Neprilysin is a member of a family of endopeptidases that includes neprilysin-like-endopeptidase, (NEPLP) α, β, and γ, PEX, XCE/DINE, and angiotensin-converting enzyme (ACE).[25,35] This molecular redundancy could have compensated for the absence of neprilysin alone. Consistently, we observed the presence of thiorphan-sensitive non-neprilysin Aβ-degrading activities in the raft (see Chapters 8 and 9) fractions from mouse brains (Takaki Y et al, submitted for publication).

2. The injection paradigm used in our initial study is accompanied by various surgical stresses because a cannula must be placed in the parenchyma for four weeks. These stresses combined with the neprilysin inhibitor infusion may have contributed the pathological states in the brain. We actually observed astrocyte activation within two days after injection of the vehicle with or without the inhibitor (Hama and Saido, unpublished data).

3. The experimental paradigm requiring long-term insertion of cannula into the brain parenchyma is likely to have caused at least slight leakage of plasma from the circulatory system directly into the brain parenchyma. Possible presence of plasma-derived amyloid-forming proteins such as amyloid P component in an unphysiological manner could have contributed to the pathological Aβ deposition.

In any case, our conclusion was that the initial approaches accompanying surgical procedures were, though more relevant than the solely in vitro paradigms, less appropriate than the reverse genetic approaches in understanding the mechanism of Aβ catabolism, particularly on a long-term basis. We have also been crossing the neprilysin-KO mice with the APP-overexpressing transgenic mice. The results seem to vary depending on which transgenic mouse strains are used. In one case, we observed that neprilysin deficiency caused more increases in soluble Aβ than insoluble Aβ (the differences were significant in both cases), in contrast to what we observed with the endogenous Aβ in nontransgenic mice; in another case, we observed a significant but lesser elevation of the Aβ levels (Iwata and Saido, unpublished

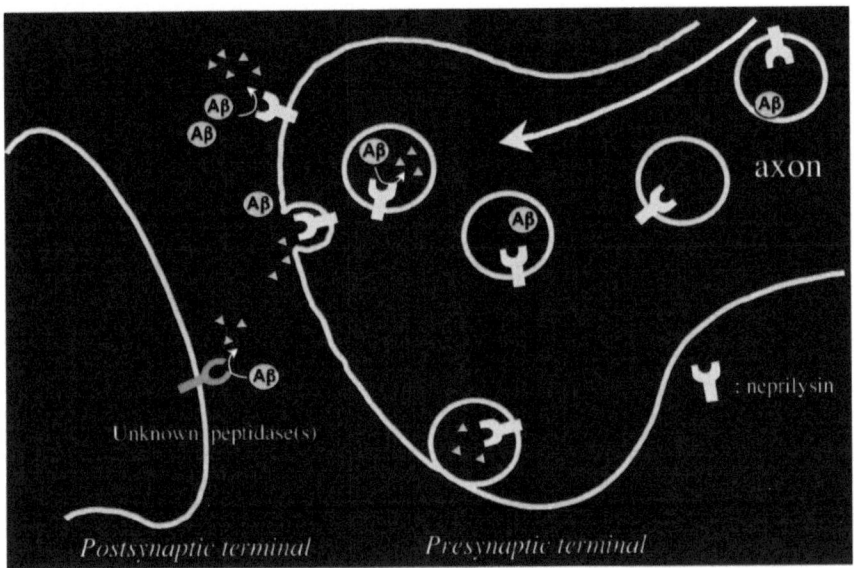

Figure 6.4. Cellular localization of neprilysin in the neuron.
Neprilysin is essentially exclusively expressed in neurons, not in glia. The peptidase, after synthesis in the soma, is axonally transported to presynaptic terminals presumably in a manner similar to the way APP is transported.

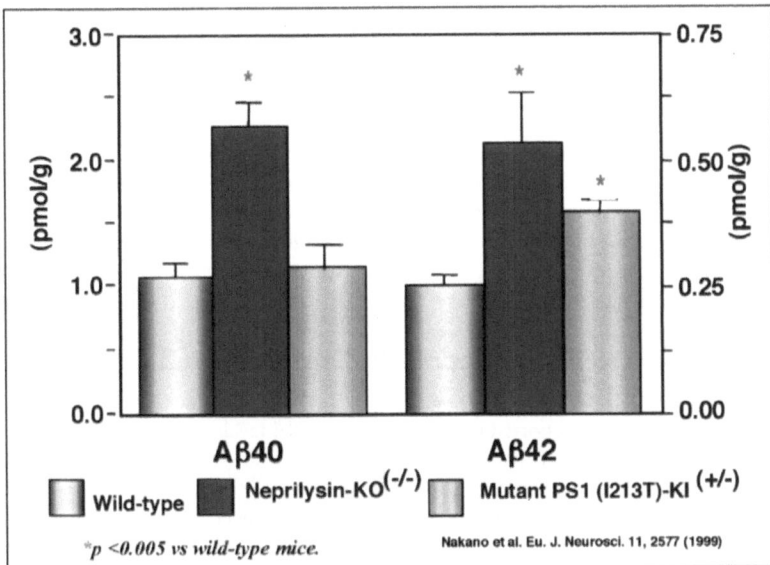

Figure 6.5. Effect of neprilysin deficiency on Aβ levels in the brain.
In neprilysin-KO mouse brains, both the endogenous $A\beta_{40}$ and $A\beta_{42}$ levels were elevated approximately 2-fold, in a manner comparable to or even greater than what has been described in FAD-causing mutant presenilin transgenic or knock-in mice. The data indicate that neprilysin alone accounts for approximately 50% of the sum of $(K_2 + K_3)$ in Formula 1.1 in Chapter 1.

Table 6.2. Effects of deficiency of various Aβ-degrading enzymes on brain Aβ levels

KO or KI Mice	$Aβ_{42}$ (Primarily Pathogenic)	$Aβ_{40}$ (Secondarily Pathogenic)
Neprilysin-KO*1	2-fold increase	2-fold increase
ECE 2-KO*2	1.2-fold increase	1.3-fold increase
IDE-KO 3	No significance*	1.1-1.2-fold increase
tPA-KO*4	No significance	No significance
uPA-KO*5	No significance	No significance
FAD presenilin 1-KI*6 (positive control)	typically 1.5-fold increase	No significance

KO: (gene) knock-out; KI: (gene) knock-in; ECE: endothelin-converting enzyme; IDE: insulin-degrading enzyme; tPA: tissue-type plasminogen activator; uPA: urokinase. *No significance if the difference is < 10%. *1: see reference 2; *2: presented by the group of Chris Eckman, Mayo Clinic, Jacksonville, at the 32nd Society for Neuroscience Meeting (November 2-7, 2002, Orlando, USA) (Heterozygous ECE 1-KO mice were also shown to increase both Aβ 42 and 40 levels 1.2-fold in the brain.); *3: presented by the group of Dennis Selkoe, Harvard Medical School, at the International ADRD Meeting (July 20-25, 2002, Stockholm, Sweden): *4: our unpublished data (Iwata and Saido); *5: presented by the group of Steve Younkin, Mayo Clinic Jacksonville, at the International ADRD Meeting (July 20-25, 2002, Stockholm, Sweden). *6: See, for instance, references 1 and 46. The quantification was performed using an identical ELISA developed by Nobuhiro Suzuki.[71] The mice were around 8-10 weeks of age. Gender is likely to be male in each study. TCS accepts the responsibility if any of the data are in error.

data). Although not proven in any relevant way, the following serve as possible explanations for the differences in the effect of neprilysin deficiency in the transgenic mice (see also Fig. 6.4).

1. Depending on the promoter used to drive the expression of transgenic APP, the degree of colocalization of the transgene-derived Aβ and neprilysin varies in a significant manner.
2. An unphysiologically high overexpression of APP (i.e., 10-100-fold) particularly in soma may alter the way Aβ is metabolized, for instance, through the endosomal-lysosomal pathway instead of being transported via secretory vesicles to presynaptic terminals.
3. Depending on the mouse strains and the cell types that express the transgene-derived APP, the transport of Aβ across the BBB or via the CSF to the circulatory system may vary to significant extent.

Region-Specific Decline of Neprilysin Expression and Activity in Brain Upon Aging

McGeer and colleagues reported that neprilysin mRNA levels are significantly reduced in the brain areas vulnerable to Aβ pathology in SAD patients at a relatively early stage (Braak stage II) as compared to the age-matched normal controls.[47,48] These observations are consistent with our hypothesis described in Section 6.1. In this respect, Aβ resembles the garbage that needs to be constantly disposed of on a daily basis (see Fig. 6.7). Just imagine the situation if the public garbage collection service for large cities like New York or Tokyo was stopped for several months; the entire city would be burdened by extremely large amount of garbage and would lose the major city functions. In our view, AD brains resemble cities in such an analogy. This analogy actually also applies to the transport-dependent clearance mechanisms (Chapters 10-12).

Nevertheless, it still remained unclear till 2002 whether reduction of neprilysin expression or activity precedes Aβ pathology during the course of aging. We thus established two

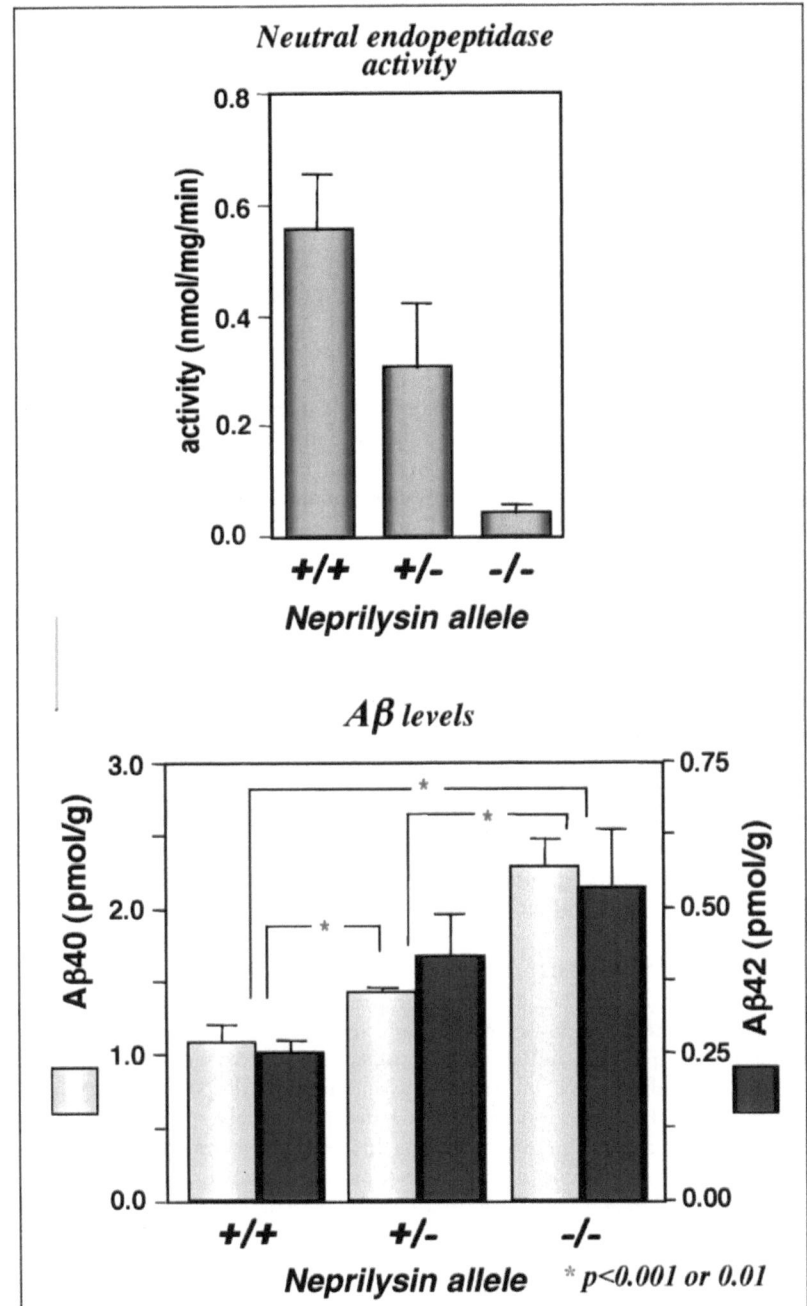

Figure 6.6. Inverse correlation between neprilysin activity and Aβ levels in the brain.
The elevation of endogenous Aβ levels was inversely correlated with the gene dose of neprilysin and thus with the enzyme activity. See ref. 2 for more details.

distinct methods to quantify the expression/activity of neprilysin in a relevant manner using neprilysin-KO mice as negative controls: (1) A biochemical assay to measure neutral endopeptidase activity in the brain (The advantage is that it is more quantitative than the latter, whereas the disadvantage is that the method fails to provide information regarding the local or spatial activity/expression of neprilysin); (2) An immunofluorescence detection using an anti-neprilysin antibody[40] (see also the previous section).

Representative immunofluorescence results are shown in Figure 6.8. The selective reduction of neprilysin expression upon aging is apparent in the polymorphic cellular, inner molecular layer, and outer molecular layer of the dentate gyrus and also in the stratum lucidum of the hippocampal CA3 sector.[49] The differences were also proven to be statistically significant according to quantitative image analyses. The results of the enzymatic quantification were also consistent with these observations; we observed a statistically significant reduction of approximately 10% of the enzyme activity per year in the entire hippocampus for two years and a 10% reduction in the neocortex in two years. If the 1.5-fold increase of $Aβ_{42}$, caused by the pathogenic mutations in the presenilin gene is truly sufficient for the development of early-onset FAD, then a 1% reduction of neprilysin activity per year, leading to 50% reduction of neprilysin activity at the age of 50 and thus about 1.5-fold elevation of the Aβ levels in the brain, would be sufficient to be causative of late-onset AD in humans. Obviously, this observation needs to be confirmed in human brains as well, but the problem is that almost all the proteins begin to be degraded within 30 min post-mortem and it is practically impossible to obtain samples prior to protein degradation in clinical situations. We therefore must indicate that proteomics using post-mortem human materials would make only a little sense if any. For reasons that we do not know, mRNAs are much more stable than proteins under such conditions. It probably is better to confirm the chronology using animals that show aging-associated Aβ amyloidosis, such as polar bears, monkeys, and dogs.[50-52]

The major implications of the immunofluorescence observations are schematized in Figure 6.9. The most important of these is that the areas where we see selective reduction of neprilysin expression correspond to the terminal zones of mossy fiber and of perforant path, which suggests that the local Aβ concentrations are particularly elevated at the presynaptic locations originally projecting from the entorhinal cortex. Note that entorhinal cortex is the region where the initial neurodegeneration takes in AD brains.

Utilization of Neprilysin Activity for Prevention and Therapy of AD

We have thus far described the possible role of neprilysin in the etiology of SAD. Our findings also indicate that regulation of neprilysin activity in a manner specific to brain regions vulnerable to Aβ deposition can be one of the strategies to reduce the Aβ burdens in the brain. Table 6.3 lists the advantages of utilizing neprilysin activity for the purpose of regulating brain Aβ levels. A relatively straight-forward approach in experimental terms, but not necessarily in clinical terms, is an application to gene therapy. Indeed, we demonstrated in an in vitro paradigm that overexpression of neprilysin, but not of an inactivated mutant form, in primary cultured neurons by Sindbis Virus leads to clearance of both the extracellular and cell-associated $Aβ_{40}$ and $Aβ_{42}$.[53]

This observation by itself however is a matter of course as has been discussed using the "lion and penguin" metaphor. Besides, it will be many years from now, if ever, when gene therapy strategy becomes clinically realistic in the treatment of most of neurological disorders. Instead, we have rather been focusing on the possible presence of specific peptide ligand-receptor system(s) that regulates the neprilysin activity/expression in neurons in a selective manner as shown in Figure 6.10. There are four major reasons upon which this assumption is based.

1. Neprilysin expression is transcriptionally regulated in a tissue specific manner.[54]

Figure 6.7. Analogy of Aβ being similar to garbage requiring daily disposal.
The pictures show a garbage truck in Tokyo and the workers as part of the public service. Just imagine the situation if the public garbage collection service for large cities was stopped for several months; the whole city would over-burdened by excessive amounts of garbage, and it would not be possible for the city to function normally. The AD brains resemble cities under such an analogy.

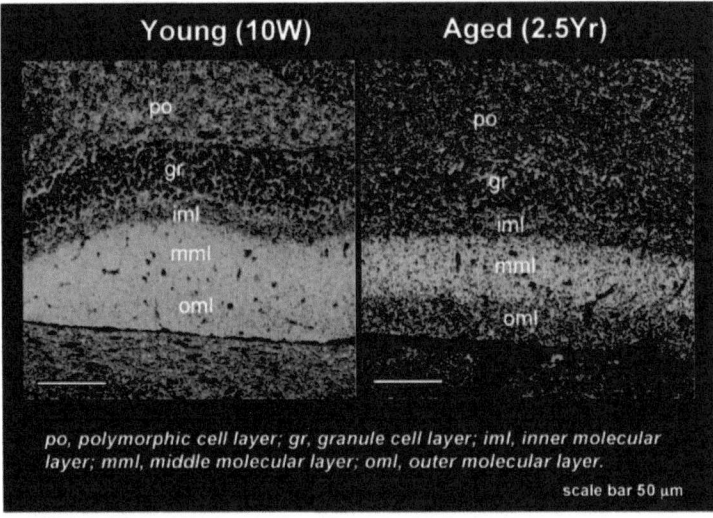

Figure 6.8. Aging-dependent reduction of neprilysin in polymorphic cell layer, inner molecular layer, and outer molecular layer in dentate gyrus.
The selective reduction of neprilysin expression is apparent in polymorphic cellular, inner molecular layer, and outer molecular layer in dentate gyrus. See ref. 49 for further details.

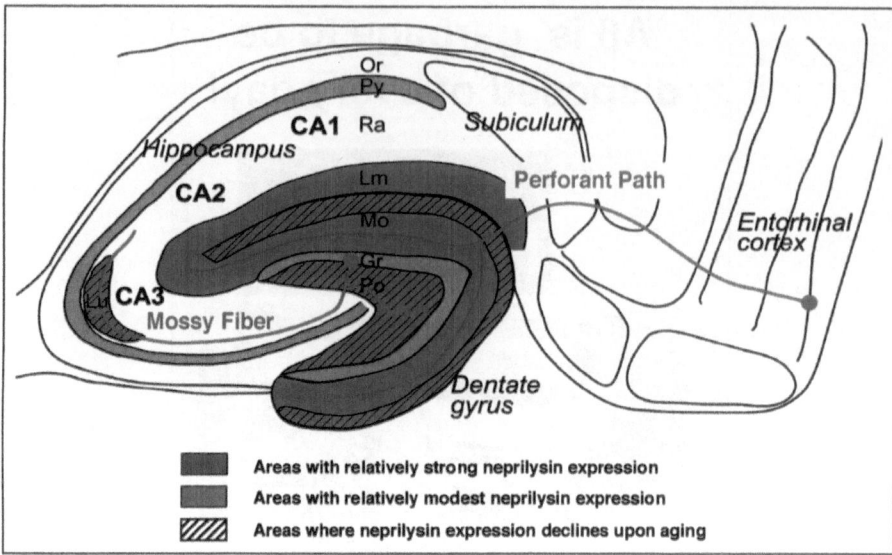

Figure 6.9. Scheme demonstrating the relationship of entorhinal cortex with the areas expressing neprilysin and those with apparent aging-dependent declines in neprilysin levels.

The hippocampal areas where we see selective reduction of neprilysin expression correspond to the terminal zones of mossy fiber and of perforant path, which suggests that the local Aβ concentrations are particularly elevated at the presynaptic locations originally projected from the entorhinal cortex. See ref. 49 for more details.

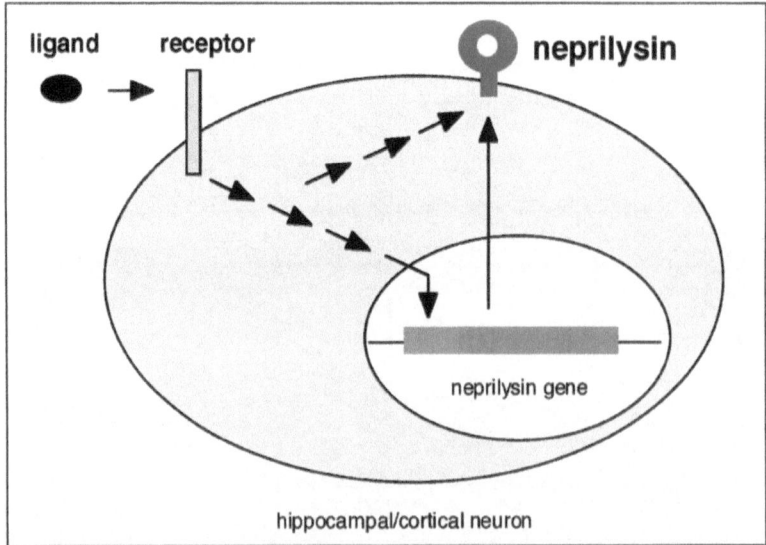

Figure 6.10. Hypothesis: Regulation of neuronal neprilysin expression/activity by a peptide ligand-system(s). We predict the presence of specific peptide ligand-receptor system(s) that regulates the neprilysin activity/expression in neurons in a selective manner. Identification of such a ligand-receptor system specifically present in the hippocampus and neocortex would provide us with an idealistic pharmacological target. The receptor is likely to be a GPCR(s).

Table 6.3. Advantages of utilizing neprilysin activity for clearance of Aβ from brain

1. Use of existing mechanism
2. No influence on processing of APP and other secretase substrates
3. Nondestructive nature (substrates < 5 kDa)
4. Preferential proteolysis of abundant substrates, i.e., Aβ in AD brain
5. Direct action on both extracellular and intracellular Aβ
6. No requirement for activation or any cofactors
7. Availability of neprilysin inhibitors for controlling hyperactivity

2. The mechanisms schematized in Figure 6.10. have been shown to exist in neutrophils and fibroblasts.[55,56] The ligands are morphine and substance P, respectively, indicating the possible presence of a specific peptide ligand(s)-regulated G-protein-coupled receptor(s) (GPCR) that may be involved in the regulation of neprilysin activity/expression in neurons.

3. Several neuropeptides have been repeatedly reported to be down-regulated in AD brains.[57-61] Some of them are also shown to be down-regulated upon aging.[62] If any one of these happens to be the ligand, the aging-dependent decline of neprilysin activity resulting in elevation of Aβ levels in the brain can be accounted for by the scheme shown in Figure 6.10.

4. Neprilysin expression is closely associated with interneurons,[40] which are the major sources of neuropeptides in the hippocampus and neocortex.[63]

Identification of such a ligand-receptor system specifically present in hippocampus and neocortex would provide us with an idealistic pharmacological target. See our forthcoming publications for further discussions and details. Note that in the history of pharmaceutical sciences approximately 60% of cases of successful utilization of clinically effective relatively small molecules involve synthetic agonists or antagonists to such receptors. Enzyme inhibitors account for about 30%. Thus, all the other approaches, which may be interesting in experimental terms, account only for about 10%. On this basis, GPCRs would be pragmatic targets.

Other Aβ-Degrading Enzyme Candidates

As shown in Table 6.1, there have been a number of other Aβ-degrading enzyme candidates thus far described mostly by in vitro paradigms. Endothelin-converting enzyme (ECE) is similar to neprilysin in structural and biochemical properties (Fig. 6.11) and indeed seems to play some role in distinct cellular compartments (see Chapter 7 for details).

Among the enzyme candidates, the most extensively studied one is insulin-degrading enzyme (IDE). IDE has an interesting characteristic in that it has the tendency of degrading potentially amyloidogenic peptides. The enzyme is essentially cytoplasmic, but only a small part of the entire population has been shown to be present on the cell surface and in the extracellular space.[64] The enzyme has also been shown to degrade the counterpart of the APP fragment (AICD) corresponding to the Notch intracellular domain (NICD).[45] One of the difficulties in studying IDE is the lack of a specific inhibitor. Insulin is often used as a competitive inhibitor but the problem is that insulin influences a number of cellular processes through activation of its receptor in cell culture and in vivo. The genome region containing the IDE gene has been claimed to be a candidate locus associated with the AD risk.[66]

Interestingly, not only neprilysin but a number of other Aβ-degrading enzyme candidates including IDE (see Table 6.1 and Fig. 6.11) are zinc-requiring enzymes. Besides, all the α-secretase candidates, a disintegrin and metalloproteinase (ADAMs) 9, 10, and 17 (see Chapters 3 and 5), which would contribute to reduction of Aβ synthesis, also require zinc for their proteolytic activities. Therefore, although too much zinc would be obviously harmful,[67] a heightened zinc deficiency is likely to be a negative risk factor for AD.

Human Genetics of Risk and Anti-Risk Factors for AD: A Proposal for Multi-Component Analysis*

In the past, researchers including ourselves tended to be more concerned about human genetic evidence for the possible involvement of neprilysin activity in AD pathogenesis than they are now. This is because the aging-dependent decline of neprilysin activity in the brain, leading to local elevation of Aβ levels in areas relevant in pathogenic terms, has been shown to be a natural process prior to any apparent AD pathology at least in mice,[49] and is, in our view, even programmed biologically (see forthcoming publications from our laboratory). Consistently, the association of neprilysin gene with the risk of developing AD does not seem to be very strong at least in the classical linkage analysis.

We have however been wondering why the apolipoprotein E ε4 allele is the only fully confirmable risk factor for SAD after hundreds or more of publications devoted to the human genetics of AD. There must be a reason. Although we are no specialists of human genetics, probably much worse than graduate students who have taken courses focusing mainly on the latest post-genome human genetics, there seems to be some limitation in the currently used techniques particularly in the search for risk and anti-risk factors for late-onset SAD. A simple question that led us to raise this issue is as follows: "Why has no one found risk or anti-risk single nucleotide polymorphisms (SNPs) or mutations in the promoter or promoter-associated regions of the APP gene in a consistent manner if the Aβ hypothesis is correct and if the presenile Aβ pathology in Down's syndrome brains is caused by only a 1.5-fold increase of the precursor of Aβ, APP, due to the trisomy of chromosome 21 carrying the APP gene?" (See also Formula 1.1 in Chapter 1).

If an individual happens to have both the risk and anti-risk factors with similar strengths but opposite effects on AD development within a given genome region, he or she would appear to be normal or average in human genetic terms in the presently used major approaches, i.e., (1) positional cloning through linkage analyses and (2) candidate gene-targeted approaches, which usually are case-control studies and occasionally longitudinal studies, focusing primarily on a single independent factor rather than a group of multiple genetic factors together. If such SNPs and mutations are randomly present in each family, the LOD (logarithm of odds) score in positional cloning would still be significantly high if a number of different families are employed but will probably appear smaller than they should be without taking the above mentioned possibility into account. Besides, the situation will be even more complex if there is some interaction among such risk factors, which is possible. Our prediction is that AD human genetics needs to introduce the latest mathematically refined approaches such as have been successfully employed in search for risk factors in cancer research,[68,69] led by the groups of E.S. Lander and others. (This may have actually begun in the AD research community without our knowledge).

We also predict that there probably will be a number of directly and indirectly influencing genetic factors not yet identified just because this could explain the presence of the substantial number of presenile cases that do not qualify as FAD by definition, not being autosomal dominantly inherited (See Table 1.1) and appearing like nongenetic SAD cases unless there exists a definitive environmental factor(s) specifically influencing the development of AD. Another fact that we have to keep in mind is that aging is the strongest risk factor; at the age of 85, one out of two people is affected either by AD or mild cognitive impairment (MCI) (see Chapter 1) whereas, at the age of 50, less than 0.1% of the population is affected. We would therefore look for risk factors in relatively young (< 60-65) SAD patients and for anti-risk factors in very old

---

*Hiroyuki Nakahara made the major contribution in this section.

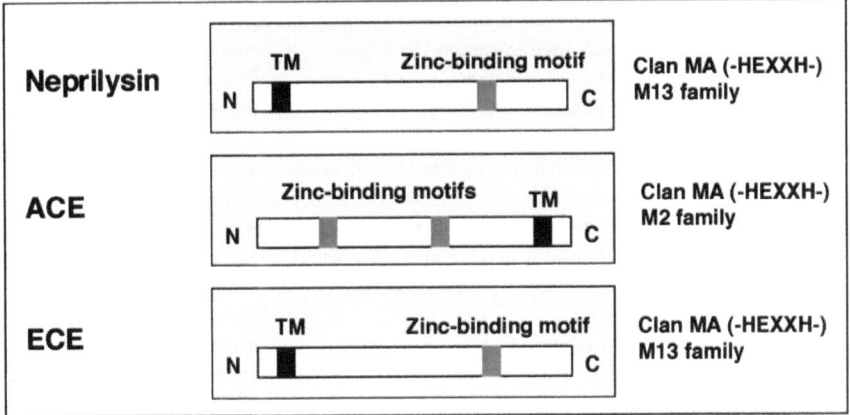

Figure 6.11. Structural similarity of other Aβ-degrading enzyme candidates to neprilysin.
Many of the candidate Aβ-degrading enzymes share some structural similarity. In particular, ACE and ECE are similar to neprilysin belonging to identical and relative clans, respectively.

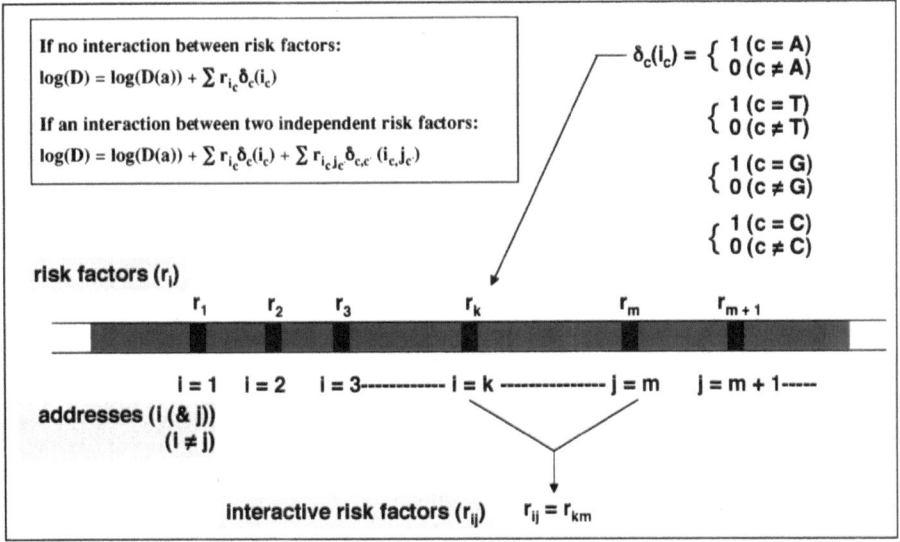

Figure 6.12. Multi-component analysis for identification of genetics risk and anti-risk factors for AD
See the section on Human Genetics of Risk and Anti-Risk Factors for AD for details.

(> 80-100) non-AD subjects. Taking these observations together, we propose the following model as a simplest example (Fig. 6.8),

$$\log(D) = \log(D(a)) + \sum r_{i_c}\, \delta_c(i_c) \qquad \text{(Formula 6.1)}$$

where D represents the degree of dementia, D(a) is a function of age, and indicates the effect of age on dementia, and the last term in the right hand side is to indicate the effect of SNPs. For the sake of simplicity, the effect of gender is excluded here but could be easily incorporated if necessary. Also note that Formula 6.1 expresses the multiplicative effect between D(a) and $\sum r_{i_c} \delta_c(i_c)$, since Formula 6.1 is equivalent to $D = D(a) \exp\{\sum r_{i_c}\delta_c(i_c)\}$. Let us explain each term:

- D: D can be any measure used to express the degree of dementia in principle but, preferably, should reflect the degree of dementia on a linear scale. One possibility is (30 - MMSE (mini-mental state examination) score), in which the values 0 to 2 would correspond to normal people and those > 8-10 would correspond to demented patients and 30 would correspond to the terminal stage. (The MMSE values seem to be quite linear at least between 10 and 23 in clinical impressions [H. Arai, personal communication]). Another possibility is a quantity representing the pathological status, i.e., total Aβ levels, quantified images of the pathological structures, etc.. This may become possible if, in the near future, we can observe the presence of Aβ amyloidosis or NFT by novel imaging techniques.

- a: age

- D(a): D(a) indicates the average "D" value at an age "a". We can simply use a look-up table for D(a) with respect to a, or use any function of "a", where a simplest example is polynomial of "a" (with some degrees).

- i: an address allocated to each SNP of a given locus (i =1, 2, 3, ----).

- c = each nucleotide species: A, T, G, or C

- r_{i_c}: a risk constant for a given SNP nucleotide species "c" (A, T, G, or C) at the address "i." . r_{i_c} will be a positive value if the SNP is a risk factor and negative if an anti-risk factor.

- $\delta_c(i_c)$: 1 if the nucleotide species of SNP at the address "i" is "c" (A, T, G, or C) and 0 if it is not "c."

For simplicity, Formula 6.1 is given so as not to include any interaction between SNPs. If it is safer to consider the possible interactions between SNPs, we would expand Formula 6.1. as follows:

$$\log(D) = \log(D(a)) + \sum r_{i_c}\delta_c(i_c) + \sum r_{i_c j_{c'}}\delta_{c,c'}(i_c, j_{c'}) \qquad \text{(Formula 6.2)}$$

where "j" corresponds to every SNP address other than "i," and "$r_{i_c j_c}$" to the risk generated by the interaction between "i_c" and "j_c". Here, Formula 6.2 includes only the second-order interaction, but it is straightforward to include the third-order and higher order interactions, if necessary. We can also add into the above model some other features, such as specific etiological, pathological, and clinical features. In addition, we note that regarding "D" as a probability distribution of being demented at a given age with given SNPs would open further possibilities to refine the above model. .

Given a set of samples, one should estimate the values of D(a) and r_{i_c} in Formula 6.1. For example, if the number of all the SNP nucleotide species in the candidate locus is 50, the number of parameters r_{i_c}, is 50 in Formula 6.1, while the number of parameters, r_{i_c} and $r_{i_c j_c}$, is $50 + {}_{50}C_2 = 1275$ at the most in Formula 6.2. For example, there is no way to estimate, say, 50 parameters, given 10 samples, without any further assumptions. In other words, depending on the number of samples, one may need to modify the above model. Then, given a model, one must estimate the values of parameters. The SNPs that present distinctly larger or smaller "r" values than most of the others tested will be the major risk and anti-risk factors, respectively. For this purpose, one must examine the relevance of estimated values of parameters and also must validate a predictive power of the model using different data sets. Furthermore, it would be more thorough if one were to examine the predictive power of different models and select the best one.

In any case, these formula are consistent with our prediction that the risk factors should be easier to identify in young patients whereas the anti-risk factors should be easier to identify in very old, normal individuals because the contribution of D(a) becomes extremely large at ages > 80-100, and thus even the strongest risk constants will be negligibly small. The effect of the apolipoprotein E ε4 allele on the incidence of AD is a good example; its effect becomes nonsignificant at ages > 80. If we were a human geneticist ourselves, we would first apply this approach to look for risk and anti-risk factors in the noncoding regions of the APP gene. In this post-genome era when more and more SNPs are being identfied, collaborations between human geneticists from all over the world may eventually make it possible to determine all the major genetic risk and anti-risk factors of AD, the knowledge of which would be very valuable not only for diagnosis but also for development of more preventive and therapeutic strategies. Remember that much of what seemed impossible even to professionals ten years ago are easily practiced now even by amateurs.

Acknowledgements

We thank all the present and past members of the Laboratory for Proteolytic Neuroscience, RIKEN Brain Science Institute, for their participation in the projects described in the first six sections. In particular, Nobuhisa Iwata and Satoshi Tsubuki have made indispensable contributions to the series of works from the beginning of the laboratory. We express our gratitude to Shun-ichi Amari (RIKEN Brain Science Institute) for valuable discussions on Mathematics and Takeshi Iwatsubo (University of Tokyo) and Hiroyuki Arai (University of Tohoku) for providing us with clinical information. We also thank our collaborators that include Maho Morishima-Kawashima and Yasuo Ihara (University of Tokyo School of Medicine), Craig Gerard and his colleagues (Harvard Medical School), Kei Maruyama (Saitama Medical School), Masatoshi Takeda and his colleagues (University of Osaka School of Medicine), Steve Younkin (Mayo Clinics, Jacksonville), John Trojanowski and Virginia Lee (University of Pennsylvania), Greg Cole (UCLA), and Peter St George-Hyslop (University of Toronto). TCS also thanks his secretary/technical assistant, Yukiko, not only for thoroughly helping him edit the entire book but also for kindly pushing him to finish everything necessary. TCS also thanks his ex-wife and still a friend, Shoko Ikuta, and his son, Ray Ikuta, for having motivated him to shift from basic biology to the more disease-oriented research through letting him know how precious one life is. Finally and most importantly, TCS thanks his wife, Tomoko, for supporting him in many ways particularly during the past difficult time he suffered with health problems after he had decided to edit this book. Our works have been supported by research grants from RIKEN BSI, Ministry of Education, Culture, Sports, Science and Technology, and Ministry of Health, Labor and Welfare of Japan, and also by a personal donation from Shigeru Sawada, whose mother died of Alzheimer's disease.

References

1. Iwata N, Tsubuki S, Takaki Y et al. Identification of the major Aβ1-42-degrading catabolic pathway in brain parenchyma: suppression leads to biochemical and pathological deposition. Nat Med 2000; 6:143-150.
2. Iwata N, Tsubuki S, Takaki Y et al. Metabolic regulation of brain Aβ by neprilysin. Science 2001; 292:1550-1552.
3. Vanderstichele H, Van Kerschaver E, Hesse C et al. Standardization of measurement of β-amyloid(1-42) in cerebrospinal fluid and plasma. Amyloid 2000; 7(4):245-258.
4. Scheuner D, Eckman C, Jensen M et al. Secreted amyloid β-protein similar to that in the senile plaques of Alzheimer's disease is increased in vivo by the presenilin 1 and 2 and APP mutations linked to familial Alzheimer's disease. Nature Med 1996; 2:864-870.
5. Hardy J, Hutton M, Farrer M et al. The genetic causes of the tau and synucleinopathies. Neurosci News 2000; 3:21-27.
6. DeMattos RB, Bales KR, Cummins DJ et al. Peripheral anti-Aβ antibody alters CNS and plasma Aβ clearance and decreases brain Aβ burden in a mouse model of Alzheimer's disease. Proc Natl Acad Sci USA 2001; 98: 8850-8855.
7. DeMattos RB, Bales KR, Cummins DJ et al. Brain to plasma amyloid-β efflux: a measure of brain amyloid burden in a mouse model of Alzheimer's disease. Science 2002; 295: 2264-2267.
8. Saido TC. Degradation of amyloid-beta peptide: a key to Alzheimer pathogenesis, prevention and therapy. Neuroscience News 2000; 3(5):52-62.

9. Sekine-Aizawa Y, Hama E, Watanabe K et al. Matrix metalloproteinase (MMP) System in Brain: Identification and Characterization of Brain-Specific MMP Highly Expressed in Cerebellum. Eur J Neurosci 2001; 13(5):935-948.

10. Akiyama H, Mori H, Saido TC et al. Occurrence of the diffuse amyloid β-protein (Aβ) deposits with numerous Aβ-containing glial cells in the cerebral cortex of patients with Alzheimer's disease. Glia 1999; 25(4):324-331.

11. Yamaguchi H, Sugihara S, Ogawa A et al. Diffuse plaques associated with astroglial amyloid β protein, possibly showing a disappearing stage of senile plaques. Acta Neuropathol (Berl) 1998; 95(3):217-222.

12. McDermott JR, Gibson AM. Degradation of Alzheimer's β-amyloid protein by human cathepsin D. Neuroreport 1996; 7:2163-2166.

13. Hamazaki H. Cathepsin D is involved in the clearance of Alzheimer's β-amyloid protein. FEBS Lett 1996; 396:139-142.

14. Kurochkin IV, Goto S. Alzheimer's β-amyloid peptide specifically interacts with and is degraded by insulin degrading enzyme. FEBS Lett 1994; 345:33-37.

15. McDermott JR, Gibson AM. Degradation of Alzheimer's β-amyloid protein by human and rat brain peptidases: involvement of insulin-degrading enzyme. Neurochem Res 1997; 22:49-56.

16. Qiu WQ, Walsh DM, Ye Z et al. Insulin-degrading enzyme regulates extracellular levels of amyloid β-protein by degradation. J Biol Chem 1998; 273:32730-32738.

17. Vekrellis K, Ye Z, Qiu WQ et al. Neurons regulate extracellular levels of amyloid β-protein via proteolysis by insulin-degrading enzyme. J Neurosci 2000; 20:1657-1665.

18. Chesneau V, Vekrellis K, Rosner MR et al. Purified recombinant insulin-degrading enzyme degrades amyloid β-protein but does not promote its oligomerization. Biochem J 2000; 351:509-516.

19. Mukherjee A, Song E, Kihiko-Ehmann M et al. Insulysin hydrolyzes amyloid β peptides to products that are neither neurotoxic nor deposit on amyloid plaques. J Neurosci 2000; 20:8745-8749.

20. Van Nostrand WE, Porter M. Plasmin cleavage of the amyloid β-protein: alteration of secondary structure and stimulation of tissue plasminogen activator activity. Biochemistry 1999; 38:11570-11576.

21. Tucker HM, Kihiko M, Caldwell JN et al. J Neurosci 2000; 20:3937-3946.

22. Exley C, Korchazhkina OV. Neuroreport 2001; 12:2967-2970.

23. Howell S, Nalbantoglu J, Crine P. Neutral endopeptidase can hydrolyze β-amyloid(1-40) but shows no effect on β-amyloid precursor protein metabolism. Peptides 1995; 16:647-652.

24. Eckman EA, Reed DK, Eckman CB. Degradation of the Alzheimer's amyloid β- peptide by endothelin-converting enzyme. J Biol Chem 2001; 276:24540-24548.

25. Hu J, Igarashi A, Kamata M et al. J Biol Chem 2001; 276:47863-47868.

26. Miyazaki K, Hasegawa M, Funahashi K et al. A metalloproteinase inhibitor domain in Alzheimer amyloid protein precursor. Nature 1993; 362:839-841.

27. Roher AE, Kasunic TC, Woods AS et al. Proteolysis of Aβ peptide from Alzheimer disease brain by gelatinase A. Biochem Biophys Res Commun 1994; 205:1755-1761.

28. Yamada T, Miyazaki K, Koshikawa N et al. Selective localization of gelatinase A, an enzyme degrading β-amyloid protein, in white matter microglia and in Schwann cells. Acta Neuropathol 1995; 89:199-203.

29. Backstrom JR, Lim GP, Cullen MJ et al. Matrix metalloproteinase-9 (MMP-9) is synthesized in neurons of the human hippocampus and is capable of degrading the amyloid-β peptide (1-40). J Neurosci 1996; 16:7910-7919.

30. Saporito-Irwin SM, Van Nostrand W. Coagulation factor XIa cleaves the RHDS sequence and abolishes the cell adhesive properties of the amyloid β-protein. J Biol Chem 1995; 270:26265-26269.

31. Hamazaki H. Carboxy-terminal truncation of long-tailed amyloid β-peptide is inhibited by serine protease inhibitor and peptide aldehyde. FEBS Lett 1998; 424:136-138.

32. Carvalho KM, Franca MS, Camarao GC et al. A new brain metalloendopeptidase which degrades the Alzheimer β-amyloid 1-40 peptide producing soluble fragments without neurotoxic effects. Braz J Med Biol Res 1997; 30:1153-1156.

33. Yamin R, Malgeri EG, Sloane JA et al. Metalloendopeptidase EC 3.4.24.15 is necessary for Alzheimer's amyloid-β peptide degradation. J Biol Che 1999; 274:18777-18784.

34. Takaki Y, Iwata N, Tsubuki S et al. Biochemical identification of the neutral endopeptidase family member responsible for the catabolism of amyloid β peptide in the brain. J Biochem (Tokyo) 2000; 128:897-902.

35. Shirotani K, Tsubuki S, Iwata N et al. Neprilysin degrades both amyloid β peptides 1-40 and 1-42 most rapidly and efficiently among thiorphan- and phosphoramidon-sensitive endopeptidases. J Biol Chem 2001; 276:21895-21901.

36. Roques BP, Noble F, Dauge V et al. Neutral endopeptidase 24.11: structure, inhibition, and experimental and clinical pharmacology. Pharmacol Rev 1993; 45:87-146.

37. Turner AJ. Neprilysin. In: Barrett AJ, Rawlings ND, Woessner JF, eds. Handbook of Proteolytic Enzymes. London: Academic Press, 1998:1080-1085.

38. Turner AJ, Isaac RE, Coates D. The neprilysin (NEP) family of zinc metalloendopeptidases: genomics and function. Bioessays 2001; 23(3):261-269.

39. Carson JA, Turner AJ. β-Amyloid catabolism: roles for neprilysin (NEP) and other metallopeptidases? J Neurochem 2002; 81(1):1-8.

40. Fukami S, Watanabe K, Iwata N et al. Aβ-degrading endopeptidase, neprilysin, in mouse brain: Synaptic and axonal localization inversely correlating with Aβ pathology. Neurosci Res 2002; 43:39-56.

41. Kiryu-Seo S, Sasaki M, Yokohama H et al. Damage-induced neuronal endopeptidase (DINE) is a unique metallopeptidase expressed in response to neuronal damage and activates superoxide scavengers. Proc Natl Acad Sci USA 2000; 97; 4345-4350.

42. Lu B, Gerard NP, Kolakowski Jr LF et al. Neutral endopeptidase modulation of septic shock. J Exp Med 1995; 181:2271-2275.

43. Duff K, Eckman C, Zehr C et al. Increased amyloid-β42(43) in brains of mice expressing mutant presenilin 1. Nature 1996; 383:710-713.

44. Borchelt DR, Thinakaran G, Eckman CB et al. Familial Alzheimer's disease-linked presenilin 1 variants elevate Aβ1-42/1-40 ratio in vitro and in vivo. Neuron 1996; 17:1005-1013.

45. Oyama F, Sawamura N, Kobayashi K et al. Mutant presenilin 2 transgenic mouse: effect on an age-dependent increase of amyloid β-protein 42 in the brain. J Neurochem 1998; 71(1):313-322.

46. Nakano Y, Kondoh G, Kudo T et al. Accumulation of murine amyloidβ42 in a gene-dosage-dependent manner in PS1 'knock-in' mice. Eur J Neurosci 1999; 11:2577-2581.

47. Yasojima K, Akiyama H, McGeer EG et al. Reduced neprilysin in high plaque areas of Alzheimer brain: a possible relationship to deficient degradation of β-amyloid peptide. Neurosci Lett 2001; 297:97-100.

48. Yasojima K, McGeer EG, McGeer PL. Relationship between β amyloid peptide generating molecules and neprilysin in Alzheimer disease and normal brain. Brain Res 2001; 919:115-121.

49. Iwata N, Takaki Y, Fukami S, et al. Region-specific reduction of Aβ-degrading endopeptidase, neprilysin, in mouse hippocampus upon aging. J Neurosci Res 2002; 70:493-500.

50. Tekirian TL, Saido TC, Markesbery WR et al. N-terminal heterogeneity of parenchymal and cerebrovascular Aβ deposits. J Neuropathol Exp Neurol 1998; 57(1):76-94.

51. Satou T, Cummings BJ, Head E et al. The progression of β-amyloid deposition in the frontal cortex of the aged canine. Brain Res 1997; 774:35-43.

52. Kanemaru K, Iwatsubo T, Ihara Y. Comparable amyloid β-protein (Aβ) 42(43) and Aβ40 deposition in the aged monkey brain. Neurosci Lett 1998; 214:196-198.

53. Hama E, Shirotani K, Masumoto H et al. Clearance of extracellular and cell-associated amyloid β peptide through viral expression of neprilysin in primary neurons. J Biochem (Tokyo) 2001; 130:721-726.

54. Li C, Booze RM, Hersh LB. Tissue-specific expression of rat neutral endopeptidase (neprilysin) mRNAs. J Biol Chem 1995; 270:5723-5728.

55. Wang TL, Chang H, Hung CR et al. Morphine preconditioning attenuates neutrophil activation in rat models of myocardial infarction. Cardiovasc Res 1998; 40(3):557-563.

56. Bae SJ, Matsunaga Y, Takenaka M et al. Substance P induced preprotachykinin-A mRNA, neutral endopeptidase mRNA and substance P in cultured normal fibroblasts. Int Arch Allergy Immunol 2002; 127(4):316-321.

57. Davies P. Katzman R. Terry RD. Reduced somatostatin-like immunoreactivity in cerebral cortex from cases of Alzheimer disease and Alzheimer senile dementa. Nature 1980; 288(5788):279-280.

58. Fujiyoshi K, Suga H, Okamoto K et al. Reduction of arginine-vasopressin in the cerebral cortex in Alzheimer type senile dementia. J Neurol Neurosurg Psychiatry 1987; 50(7):929-932.

59. Beal MF, Mazurek MF, Chattha GK et al. Neuropeptide Y immunoreactivity is reduced in cerebral cortex in Alzheimer's disease. Ann Neurol 1986; 20(3):282-288.

60. Crystal HA. Davies P. Cortical substance P-like immunoreactivity in cases of Alzheimer's disease and senile dementia of the Alzheimer type. J Neurochem 1982; 38(6):1781-1784.

61. Bissette G, Reynolds GP, Kilts CD et al. Corticotropin-releasing factor-like immunoreactivity in senile dementia of the Alzheimer type. Reduced cortical and striatal concentrations. JAMA1985; 254(21):3067-3069.

62. Fliers E, Swaab DF. Neuropeptide changes in aging and Alzheimer's disease. Prog Brain Res 1986;70:141-152.

63. Freund TF, Buzsaki G. Interneurons of the hippocampus. Hippocampus 1996; 6(4):347-470.

64. Roth RA. Insulysin. In: Barrett AJ, Rawlings ND, Woessner JF, eds. Handbook of Proteolytic Enzymes. London: Academic Press, 1998:1362-1367.

65. Edbauer D, Willem M, Lammich S et al. Insulin-degrading enzyme rapidly removes the β-amyloid precursor protein intracellular domain (AICD). J Biol Chem 2002; 277(16):13389-13393.

66. Bertram L, Blacker D, Mullin K, et al. Evidence for genetic linkage of Alzheimer's disease to chromosome 10q. Science 2000; 290: 2302-2303.

67. Cherny RA, Atwood CS, Xilinas ME et al. Treatment with a copper-zinc chelator markedly and rapidly inhibits β-amyloid accumulation in Alzheimer's disease transgenic mice. Neuron 2001; 30:665-676.

68. Shipp MA, Ross, KN, Tamayo P et al. Diffuse large B-cell lymphoma outcome prediction by gene-expression profiling and supervised machine learning. Nature Med 2002; 8:68-74.

69. Tibshirani R, Hastie, T, Narasimhan B et al. Diagnosis of multiple cancer types by shrunken cntroids of gene expression. Proc Natl Acad Sci USA 2002; 99:6567-6572.

70. Lafrance MH, Vezina C, Wang Q et al. Role of glycosylation in transport and enzymic activity of neutral endopeptidase-24.11. Biochem J 1994; 302 (2):451-454.

71. Suzuki N, Cheung TT, Cai XD, et al. An increased percentage of long amyloid β protein secreted by familial amyloid β protein precursor (βAPP717) mutants. Science 1994; 264(5163):1336-1340.

CHAPTER 7

Aβ Degradation by Endothelin-Converting Enzymes

Elizabeth A. Eckman and Christopher B. Eckman

Abstract

The abnormal accumulation of β-amyloid (Aβ) in the brain is an early and invariant feature of Alzheimer's disease and is believed to play a pivotal role in the etiology and pathogenesis of the disease. The concentration of Aβ is regulated by multiple enzymatic activities, including proteases responsible for its degradation. In this chapter we present evidence for endothelin-converting enzymes (ECEs) as Aβ-degrading enzymes. Overexpression of ECE-1 in cultured cells reduces Aβ accumulation by up to 90%, and the enzyme is capable of directly cleaving Aβ at multiple sites. As ECEs are expressed in brain, reduced ECE activity by genetic mutation, altered transcriptional activity, or pharmacological inhibition, for example, may be a risk factor for Alzheimer's disease (AD). The risk of pharmacological reduction of ECE activity is of particular concern since ECE inhibitors are being developed for the treatment of hypertension and other disorders.

Introduction: The Endothelin-Converting Enzymes

The endothelin-converting enzymes (ECEs) are a class of type II integral membrane zinc metalloproteases (active site lumenal) named for their ability to hydrolyze a family of biologically inactive intermediates, big endothelins (big ETs), exclusively at a Trp^{21}-Val/Ile^{22} bond to form the potent vasoconstrictors endothelins.[1] This specific cleavage event appears to be determined in part by the secondary/tertiary structure of disulfide-bonded big ETs, as linear big ET-1 (in which the cysteines have been alkylated) is cleaved at multiple sites by ECE.[2] Targeted disruption of the ECE-1 gene in mice revealed that this enzyme is in fact a physiologically relevant activating enzyme for both ET-1 and ET-3, and ECE-1 null mice die in utero or within minutes of birth due to severe craniofacial and cardiac defects.[3] In addition to the specific cleavage of big ETs, ECE has been reported to hydrolyze several biologically active peptides in vitro, including bradykinin, neurotensin, substance P, and oxidized insulin B chain by cleaving on the amino side of hydrophobic residues.[4,5]

Two different endothelin-converting enzymes have been cloned. The first identified, ECE-1, is abundantly expressed in the vascular endothelial cells of all organs and is also widely expressed in nonvascular cells of tissues including brain, lung, pancreas, testis, ovary, and adrenal gland.[6-8] Four isoforms of human ECE-1 differing only in the cytoplasmic tail are produced by a single gene located on chromosome 1 (1p36) through the use of alternate promoters.[6,9-13] The four isoforms cleave big ETs with equal efficiency but differ primarily in their subcellular localization and tissue distribution.[12,13] Human ECE-1a is localized predominantly to the plasma membrane.[12,13] Human ECE-1c and ECE-1d have also been reported to be localized predominantly to the plasma membrane with additional intracellular expression detected.[12,13] In contrast, human ECE-1b appears to be localized exclusively intracellularly. Co-immunolocalization studies performed by Schweizer et al[12] on human ECE-1b transfected CHO cells indicate the presence of this isoform in the trans-Golgi network (TGN). Azarani et al similarly demonstrated that human ECE-1b was located in an intracellular compartment when expressed in Madin-Darby canine kidney (MDCK) cells.[14] In an endogenous ECE-1b and ECE-1c expressing cell line, ECV304, ECE-1 immunoreactivity was detected in intracellular Golgi-like structures as well as at the cell surface.[12]

ECE-2 is a homologous enzyme with catalytic activity similar to that of ECE-1. ECE-2 is localized intracellularly, has an acidic pH optimum, and is expressed most abundantly in the nervous system.[15] Immunocytochemical analysis of endogenous ECE-2 in HUVEC cells revealed a punctate pattern of staining consistent with expression of ECE-2 in acidic intracellular vesicles of the constitutive secretory pathway.[16] Like ECE-1, ECE-2 cleaves big ET-1 most efficiently among the three big ETs, and at least three isoforms of human ECE-2 are produced from a single gene on chromosome 3 (3q28-q29).[12,15,17] The catalytic activity, subcellular localization, and tissue distribution of the isoforms have not yet been compared.

ECE-2 null mice have been made and appear normal and healthy.[18] In ECE-1/ECE-2 double knockout embryos the levels of two known products of ECE activity, ET-1 and ET-2, are remarkably similar to the ECE-1 nulls alone. The highly restricted tissue distribution of ECE-2 and modest effect of disruption of the ECE-2 gene on an ECE-1 null background suggests that ECE-2 may have physiological substrates distinct from the big ETs.[18]

Both ECE-1 and ECE-2 Are Expressed in Brain

While many classes of proteases may degrade Aβ in vitro, to be physiologically relevant Aβ degrading enzymes must be expressed in the brain and their cellular localization must be consistent with having access to Aβ peptides. In the nervous system, ECE-1 immunoreactivity has been detected in fibers within the glial limitans, in neuronal processes and cell bodies of the cerebral cortex, in pyramidal cells of the neocortex and hippocampus, in astrocytes, and in Purkinje cells in the cerebellum.[7,19-21] Northern blot analysis of bovine tissues revealed that ECE-2 is most abundantly expressed in neural tissues including cerebral cortex, cerebellum, and adrenal medulla.[15] In mouse brain, ECE-2 is expressed in heterogeneous populations of neurons in the thalamus, hypothalamus, amygdala, dentate gyrus, and CA3.[18] While detailed co-localization studies have not been performed, separate studies indicate that ECE and Aβ are present in the same cellular compartments. Human ECE-1b has been reported to be present in the TGN, a proposed site of Aβ40 generation in neuronal cells.[12,22] ECE-2 is also likely to be present in the TGN and vesicles of the constitutive secretory pathway.[15,16] The topology of the ECEs is such that the active site is within the lumen of organelles and vesicles, providing access to the Aβ peptide.

A Potential Role for Endothelin-Converting Enzymes in Aβ Catabolism

The role that ECE may play in Alzheimer's disease (AD) is only beginning to be explored. While the data are controversial and factors other than ECE may contribute to endothelin levels, it is worth noting that endothelin levels have been reported to be decreased in the CSF of patients with AD when compared to nondemented control patients.[23] Sib-pair analyses of genetic factors contributing to late onset AD have not excluded the region on chromosome 1 where the ECE-1 gene is located,[24] and interestingly the human ECE-2 gene is located on chromosome 3 in the vicinity of a possible AD locus.[17,25,26] In this chapter we will present pharmacological and biochemical evidence for ECEs as Aβ degrading enzymes.

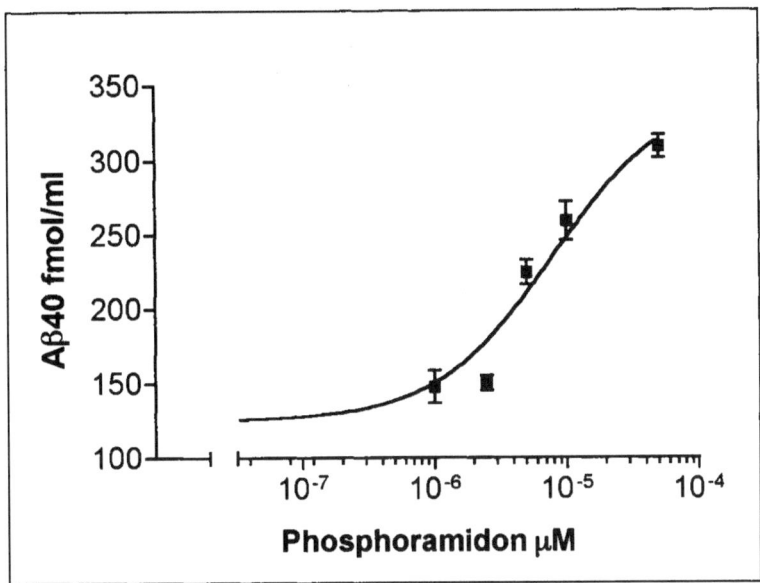

Figure 7.1. Effect of phosphoramidon on Aβ accumulation by H4 cells.
Confluent wells of H4 cells were incubated overnight with phosphoramidon at the indicated concentrations. The medium was removed and $Aβ_{40}$ concentration determined by sandwich ELISA. Data shown represents the mean ± SE of triplicate wells for each dose, each measured in duplicate.

The Serendipitous Discovery of Metalloproteases as Important Modulators of Aβ Concentration in vivo

In the mid-1990s a major focus of Alzheimer's disease research in our laboratory and others was the identification of the proteases responsible for the generation of Aβ peptides from a larger precursor, the β-amyloid precursor protein or βAPP. Two distinct proteolytic activities, termed β and γ secretase, are required for the release of Aβ from the membrane-bound βAPP, and the identification of specific inhibitors of these activities would aid greatly in the identification and characterization of the then-elusive proteases. As such, our laboratory screened an extensive collection of known protease inhibitors and other compounds for their ability to reduce Aβ secretion by a human CNS-derived cell line, H4. These screens revealed not only compounds that reduced Aβ concentration, but also some that dramatically increased the accumulation of the peptide. One of the most notable of these was the metalloprotease inhibitor phosphoramidon, which caused a 2-3 fold elevation of Aβ concentration in the medium of H4 cells (Fig. 7.1). This finding was particularly interesting to us as the effect on Aβ observed following phosphoramidon treatment was as great or greater than most of the AD-causing mutations that we and others had previously examined.[27]

To determine whether phosphoramidon could modulate Aβ concentration in vivo as well as in cell culture, we examined the effect of the compound following intravenous administration in guinea pigs, an animal model with an Aβ peptide sequence identical to that in humans. Phosphoramidon treatment caused a rapid, dose-dependent increase in the concentration of Aβ in the plasma of guinea pigs (Fig. 7.2), indicating that one or more phosphoramidon-sensitive proteases play an important role in regulating Aβ concentration in vivo.

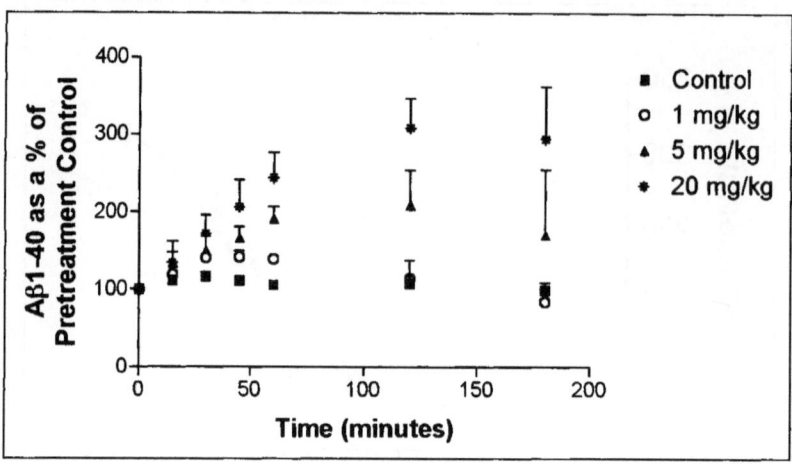

Figure 7.2. Effect of intravenous administration of phosphoramidon on plasma Aβ concentration in the guinea pig.
Phosphoramidon was administered at the indicated concentrations as an i.v. bolus. At various time points, blood was removed and plasma prepared and assayed by sandwich ELISA for $Aβ_{1-40}$. Data points represent the mean ± SE from at least two animals, each measured in duplicate.

Phosphoramidon Increases Aβ Concentration in CNS-Derived Cell Lines Through the Inhibition of Intracellular Degradation

There are multiple mechanisms by which phosphoramidon could cause an increase in Aβ concentration. For example, inhibition of α-secretase, which cleaves within the Aβ peptide, could cause an increase in the production of Aβ by increasing the amount of substrate available for the Aβ-generating β and γ secretases. This mechanism, however, was unlikely to be responsible for the phosphoramidon-mediated effect on Aβ as the levels of other α-secretase derived APP products such as secreted APP-α (sAPPα), and C-terminal fragment α (CTFα) were unchanged in treated cells.[28,29] An intriguing possibility was that phosphoramidon was somehow inhibiting the turnover of Aβ.

Proteolytic degradation of Aβ may occur at multiple sites, both intracellular and extracellular. To determine if the phosphoramidon-mediated increase in Aβ was due to inhibition of a secreted protease we first examined the amount of Aβ remaining in isolated H4 conditioned medium that was incubated at 37°C. While Aβ degradation was readily apparent, as assessed by loss of signal in our sandwich ELISAs with increasing incubation time, this degradation was not sensitive to phosphoramidon (data not shown). To examine whether the phosphoramidon induced increases in Aβ could be due to inhibition of a cell-surface protease we performed a spike experiment in which exogenous Aβ was added to the medium bathing H4 cells in the absence or presence of phosphoramidon. Exogenous Aβ was removed equally well in the presence or absence of phosphoramidon, indicating that the phosphoramidon-induced effect is not likely due to decreased internalization or to inhibition of a cell-surface or secreted protease.[29] Collectively, these data suggested that phosphoramidon was exerting its effect on an intracellular event that eventually culminated in an increase in extracellular Aβ concentration. This is consistent with a report by Fuller et al showing an increase in cell-associated Aβ in phosphoramidon-treated SY5Y cells.[28] In H4 cells, however, we have been unable to detect an elevation in intracellular Aβ levels, presumably due to rapid secretion of Aβ in this model system.

Endothelin-Converting Enzymes: Potential Targets for the Phosphoramidon-Mediated Increase in Aβ Concentration

Phosphoramidon is known to inhibit several metalloproteases including ECE-1, ECE-2, neprilysin, and angiotensin-converting enzyme (ACE), but not insulin-degrading enzyme (IDE).[30,31] Neprilysin and ACE, which are both capable of degrading Aβ, have been reported to reside predominantly on the cell surface, although a soluble form of neprilysin is also present in serum and CSF.[32-38] A recently identified phosphoramidon and thiorphan-sensitive neprilysin homologue SEP/NL1/NEPII is expressed both as a membrane-bound and secreted protease.[39-41] While our spike experiments suggested that the phosphoramidon-induced increases in Aβ are not likely due to inhibition of a cell-surface or secreted enzyme, we decided to further evaluate a role for neprilysin and ACE in our cell culture system using more selective inhibitors of these enzymes, thiorphan and captopril, respectively. We have not yet been able to similarly analyze ECE, as a more selective inhibitor of ECE is not commercially available.

Treatment of H4 cells with phosphoramidon resulted in a greater than 2-fold elevation in $A\beta_{40}$ accumulation, with a half-maximal effect occurring at a dose of approximately 7.5 μM (see Fig. 7.1). Treatment with thiorphan or captopril at concentrations greater than 1000 times the reported IC_{50} for the target enzymes in in vitro studies,[42,43] but less than that required to inhibit ECE, failed to result in increases in extracellular Aβ (Fig. 7.3). These data indicated that the phosphoramidon-induced effect in H4 cells is not likely due to inhibition of neprilysin, SEP or ACE. Similar results were obtained for $A\beta_{42}$ (data not shown). As endogenous ECE activity was readily detected in solubilized membranes of H4 cells using a big ET conversion assay[6,44] (data not shown), our focus then shifted to evaluating a role for ECE in Aβ degradation.

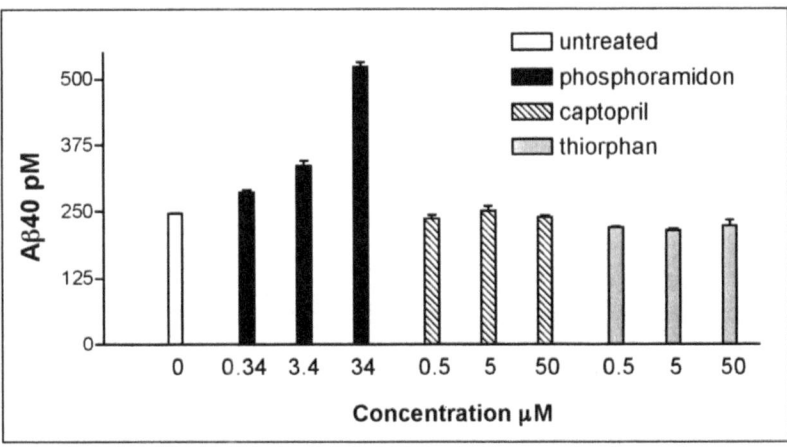

Figure 7.3. Effect of metalloprotease inhibitors on Aβ concentration in the culture medium of H4 cells. Confluent wells of H4 cells were treated with phosphoramidon, captopril (selective inhibitor of ACE), or thiorphan (selective inhibitor of neprilysin) for 24 hours at the indicated concentrations. $A\beta_{40}$ concentration in the conditioned media was determined by sandwich ELISA. Data are plotted as mean ± SE of triplicate wells. MTS assays on the treated cells did not reveal cellular toxicity at any of the doses (data not shown). Reprinted with permission from: Eckman EA, Reed DK, Eckman CB. J Biol Chem 2001; 276(27):24540-8, 2001 American Society for Biochemistry and Molecular Biology.

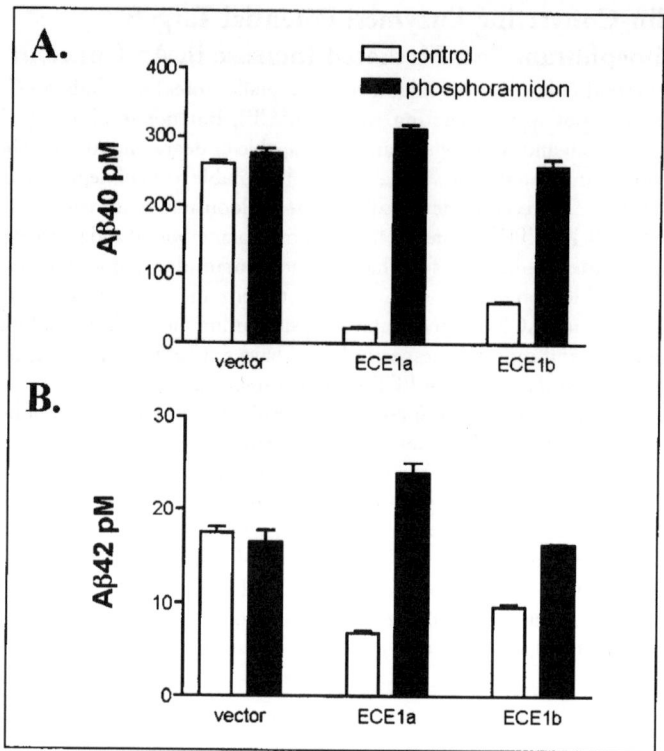

Figure 7.4. Overexpression of ECE-1a or ECE-1b in CHO cells reduces extracellular Aβ concentration. $Aβ_{40}$ (A) and $Aβ_{42}$ (B) concentration in the conditioned media of stable ECE-1a and ECE-1b transfected CHO cell lines was determined by sandwich ELISA following a 24 hour incubation ± 100 μM phosphoramidon. Data are plotted as mean ± SE of triplicate wells. Reprinted with permission from: Eckman EA, Reed DK, Eckman CB. J Biol Chem 2001; 276(27):24540-8, 2001 American Society for Biochemistry and Molecular Biology.

Overexpression of Endothelin-Converting Enzyme-1 Results in a Significant Decrease in Extracellular Aβ Concentration that Is Completely Reversed by Treatment with Phosphoramidon

Evidence implicating a potential role for ECE in modulating Aβ concentration came further from the observation that CHO cells, which have no endogenous ECE activity,[6] produce very high levels of Aβ when compared to most other cell types (personal observation). Conversely, HUVEC cells, which have high levels of endogenous ECE,[45] accumulate very little Aβ unless treated with high concentrations of phosphoramidon (data not shown). To follow up on these casual observations and to further investigate the role of ECE in Aβ accumulation we cloned and stably transfected CHO cells with human ECE-1a and ECE-1b. Overexpression of either ECE-1a or ECE-1b in CHO cells, which lack endogenous ECE activity, resulted in a striking 75-90% reduction in $Aβ_{40}$ and a 45-60% reduction in $Aβ_{42}$ (Fig. 7.4). No significant changes were observed in the amount of sAPP accumulation in ECE-transfected cells compared to the vector controls, indicating that the cells were similarly viable and that general secretion is not affected by ECE overexpression. The reduction in Aβ concentration in ECE-transfected cells was completely reversed by treatment with phosphoramidon, indicating that the observed phenotype was likely due to the enzymatic activity of the overexpressed ECE.

Increased Removal of Exogenous Aβ Is Apparent Only in ECE-1a Transfected Cells

In an endogenous ECE-expressing cell line, H4, extracellular Aβ removal does not appear to be affected by phosphoramidon treatment.[29] To determine whether extracellular Aβ removal could account, at least in part, for the dramatic decrease in extracellular Aβ concentration in ECE-transfected cell lines, we spiked Aβ into the culture medium in the presence or absence of phosphoramidon and determined the percent removal by sandwich ELISA at 6 and 24 hours. Following a 6 hour incubation, removal of Aβ was similar in the culture medium of vector and ECE-transfected CHO cells, and was not affected by phosphoramidon treatment, although secretion of endogenous Aβ in phosphoramidon treated ECE-transfected cells was increased 1.5 to 2-fold during the same time period.[29] After a 24-hour incubation, we did observe a significant increase in the removal of the spiked-in Aβ in the medium of ECE-1a transfected cells compared to the vector controls (Fig. 7.5). No significant change in exogenous Aβ removal was observed in cells expressing ECE-1b. The ECE-1a induced increase in Aβ removal could be completely attenuated by phosphoramidon treatment, indicating that the effect was likely due to the enzymatic activity of ECE-1a (Fig. 7.5, inset).

These results suggest that ECE may contribute slightly to extracellular Aβ removal, at least in cells overexpressing human ECE-1a. However, the dramatic increase in Aβ concentration in ECE-1a expressing cells upon treatment with phosphoramidon (see Fig. 7.4) does not appear to be accounted for by the modest increase in exogenous Aβ degradation. While ECE-1a has been reported to be localized predominantly to the cell surface, this isoform has been

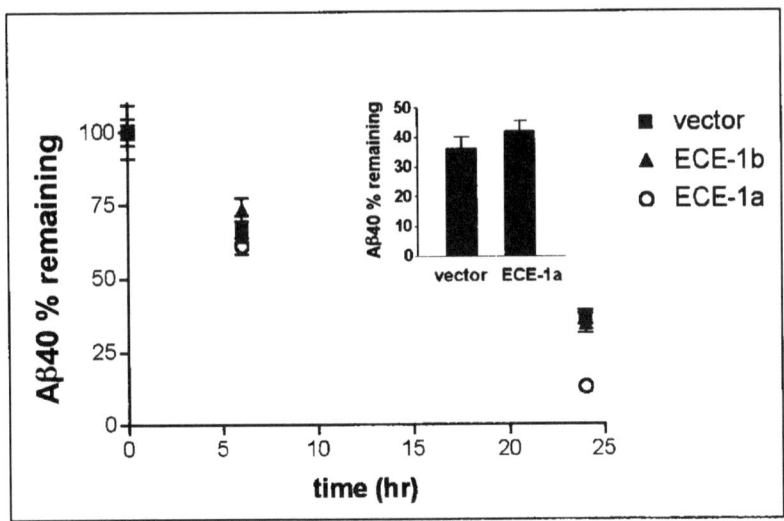

Figure 7.5. Removal of exogenous synthetic Aβ by ECE-overexpressing CHO cells.
Synthetic human Aβ$_{40}$ (150 pM) was added to confluent CHO cells stably transfected with ECE-1a, ECE-1b, or the control vector, and incubated for 6 and 24 hours. Human Aβ$_{40}$ was measured at the indicated time points using the BAN50/BA27 sandwich ELISA which does not detect endogenous CHO Aβ. Data shown represents the mean ± SE of triplicate wells that were incubated with synthetic Aβ$_{40}$. The concentration of Aβ remaining after 24 hours is significantly lower in ECE-1a cells than in vector controls (p=0.0495, Mann-Whitney). Inset graph shows percent of exogenous Aβ removed by ECE-1a and vector-transfected cells after 24 hours in the presence of phosphoramidon (34 µM). Reprinted with permission from: Eckman EA, Reed DK, Eckman CB. J Biol Chem 2001; 276(27):24540-8, 2001 American Society for Biochemistry and Molecular Biology.

shown to process big ET-1 intracellularly in CHO cells, most likely in secretory vesicles, as well as at the cell surface.[46] Therefore, ECE-1a may similarly degrade Aβ intracellularly in CHO cells as it is being trafficked to the cell surface. Ultimately, the result of increased intracellular degradation is a decrease in extracellular Aβ accumulation.

Recombinant ECE-1 Degrades Aβ in vitro

Recombinant, soluble forms of ECE-1 (solECE-1) lacking the intracellular and trans-membrane domains have been reported to hydrolyze big ET-1 with activity comparable to that of membrane-bound ECE-1a.[5,44,47] To examine whether ECE-1 is capable of direct catabolism of Aβ we generated a soluble ECE-1 similar to those previously described. Incubation of synthetic $Aβ_{40}$ and $Aβ_{42}$ with this enzyme resulted in a nearly complete loss of the full-length peptides as detected by sandwich ELISA, and this reduction was completely blocked by incubation with ECE inhibitors.[29]

ECE-1 has been shown to cleave a number of biologically active peptides on the amino side of hydrophobic residues and appears not to cleave peptides smaller than 6 amino acids in length.[5] Given this specificity, there are approximately 13 potential ECE cleavage sites in the Aβ40 peptide. Using HPLC, mass spectrometry, and NH_2-sequence analysis[29] we determined that soluble ECE-1 cleaves synthetic $Aβ_{40}$ at at least three sites, resulting in the formation of Aβ fragments 1-16, 1-17, 1-19, and 20-40 (Fig. 7.6). Consistent with the known substrate specificity of ECE-1, each of these observed cleavages by solECE-1 occurred on the amino side of hydrophobic residues (Leu[17], Val[18], Phe[20]). Given that ECE-2 is highly homologous to ECE-1 and shares similar catalytic activity,[15] ECE-2 is also likely capable of degrading Aβ.

Kinetic Analysis of $Aβ_{40}$ Cleavage by solECE-1

To determine the catalytic efficiency of Aβ cleavage by solECE-1 we calculated k_{cat}/K_m by measuring the rate of Aβ hydrolysis under second-order conditions at pH 6.5, the reported pH optimum for big ET-1 cleavage by the enzyme. Using this method, the k_{cat}/K_m for $Aβ_{40}$ hydrolysis by solECE-1 was determined to be $(1.7 +/- 0.6) \times 10^3$ $M^{-1}sec^{-1}$ at pH 6.5, a value 15-fold lower than that for big ET-1 hydrolysis under the same conditions.[29] Intrigued by a report indicating that solECE-1 cleaves bradykinin and substance P with an acidic pH optimum of approximately 5.6 compared to the optimum of pH 6.5 for big ET-1,[48] we next compared the efficiency of solECE-1 hydrolysis of $Aβ_{40}$ and big ET-1 at pH 5.6. Similar to bradykinin and substance P, solECE-1 cleaves $Aβ_{40}$ greater than an order of magnitude more efficiently at pH 5.6 than at pH 6.5. The value of k_{cat}/K_m for big ET-1 hydrolysis at pH 5.6 is only 3-fold greater than that for $Aβ_{40}$ at the same pH. This result may be particularly relevant as the TGN, where ECE-1b appears to be expressed, has an acidic pH. It may also help to explain why little exogenous Aβ is degraded by ECE-1a in transfected CHO cells.

Current Studies in Animal Models

A reasonable test for the physiological involvement of ECE or any enzyme in contributing to Aβ accumulation in the brain is to examine the effect of animals null for the enzyme. While several enzymes have been identified that can degrade Aβ in in vitro assays, only neprilysin has thus far been reported to influence Aβ accumulation in the brains of knock-out mice.[33,49,50] Ongoing studies in our laboratory are focused on determining the effect of ECE deficiency on Aβ accumulation in the brains of knockout mice. As in the neprilysin knockout mice, elevated levels of Aβ in the brains of ECE knockout mice would indicate a physiological role for this enzyme family in regulating the accumulation of the peptide. Crossing the ECE knockout mice with a mouse model that develops AD-like pathology will further test the hypothesis that reduced ECE activity may critically alter the concentration of Aβ in the brain, accelerating and/or enhancing the development of senile plaques.

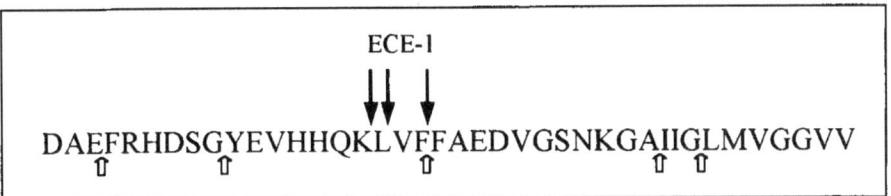

Figure 7.6. Sites of $Aβ_{40}$ cleavage by solECE-1.
Sites of $Aβ_{40}$ cleavage by solECE-1 (solid arrows) were determined by HPLC, mass spectrometry, and NH_2-sequence analysis.[29] Open arrows under the sequence indicate previously determined neprilysin cleavage sites.[32]

Summary: ECE Activity May Be One of Many Regulators of Aβ Accumulation in the Brain

It is likely that multiple proteases, both intracellular and extracellular, may play a role in determining Aβ concentration. Intracellular degradation of Aβ at the site of generation and/or within the secretory pathway may affect the extracellular concentration of the peptide by limiting the amount available for secretion. The concentration of secreted Aβ is further regulated by direct degradation by extracellular proteases and by receptor-mediated endocytosis or phagocytosis followed by lysosomal degradation. The relative contribution of Aβ degrading enzymes and other mechanisms of Aβ removal may vary in different regions of the brain and may also differ for $Aβ_{40}$ and $Aβ_{42}$. Decreases in any one of these mechanisms of Aβ removal, whether major or minor, may potentially result in increased Aβ accumulation and the development of AD pathology. Conversely, an increase in the activity of any enzyme capable of degrading Aβ may result in decreased accumulation of the peptide, potentially reducing the risk for AD.

We have presented extensive evidence that ECE is capable of degrading Aβ peptides, and studies in cell-based models suggest that the major site of ECE's effect on Aβ is intracellular, possibly in an acidic compartment. Even in cells overexpressing ECE-1a, an isoform expressed predominantly on the cell-surface, little ECE-mediated degradation of extracellular Aβ was observed. This may be due to the unfavorable kinetics of Aβ degradation by ECE at neutral pH and suggests that the dramatic reduction in Aβ concentration seen in these cells is due to degradation of Aβ within the biosynthetic, secretory, endocytic or other slightly acidic organelles and pathways. Interestingly, inhibition of this intracellular event(s) has the net result of a substantial increase in extracellular Aβ.

The Regulation of Endogenous ECE Activity Is Complex, and Alterations May Influence Susceptibility to Alzheimer's Disease

The regulation of ECE expression is complex, as evidenced by the production of multiple isoforms of both ECE-1 and ECE-2 through the use of multiple promoters and alternative splicing. ECE-1 activity may be modulated at the level of transcription, mRNA stability, glycosylation, and zinc binding (see for example refs. 51-59). ECE-1 expression is increased under a number of pathophysiological conditions including but not limited to hypertension, congestive heart failure, subarachnoid hemorrhage, preeclampsia, and wound healing.[55,60-62] In cultured endothelial cells, thrombin upregulates ECE-1 expression via the ERK pathway.[63] It has been reported that ECE-1 activity is regulated transcriptionally and/or post-transcriptionally by endothelins (down-regulated via ET_B receptors), and angiotensin-II (up-regulated via ET_A receptors).[58,64-66] Glycosylation is required for the functional expression of ECE-1[56,57] and altered glycosylation may be involved in the negative feedback regulation of ECE-1 by ET_B receptors.[58] Superoxide, which may be generated during a variety of pathophysiological conditions, reversibly inhibits ECE by causing the release of zinc from the active site.[59]

Additional studies are required to determine the extent of the role of ECE activity in determining Aβ concentration in the brain, and the possible link between altered ECE activity and the development and/or progression of Alzheimer's disease. The complex regulation of ECE activity suggests that physiological conditions that cause a reduction in ECE activity in the brain may elevate Aβ levels and increase susceptibility to AD. Alterations in ECE activity in the brain (either in total activity or in the relative expression of ECE isoforms) may possibly occur in normal aging, the most common risk factor for AD. Genetic mutations in ECE that decrease its activity may also be identified that are causative of Alzheimer's disease in certain individuals. Equally important, however, is the possibility that there may be individuals with normally high levels of ECE activity who are at a reduced risk for the disease. A careful analysis of ECE activity in AD and control individuals is necessary to determine the extent of the involvement of this enzyme family in the development of AD.

Increased ECE Activity May Be Therapeutic for Alzheimer's Disease

Regardless of the extent of the influence of endogenous ECE activity on Aβ accumulation in vivo, increased activity of ECE or any other Aβ-degrading enzyme may be beneficial in AD. Up-regulation of ECE activity through gene therapy, transcriptional activation, or by reducing ECE turnover, for example, may provide a novel therapeutic approach for the treatment of AD. One obvious concern with this treatment method is that patients may become hypertensive. Up-regulation of ECE activity in the periphery of rats by intravenous injection of an adenoviral construct containing a secreted form of ECE does not, however, appear to result in increased circulating endothelin levels or in hypertension, indicating that ECE is not likely to be rate-limiting in the biosynthesis of ET under these conditions.[67] Further, even if increased ECE activity does augment endothelin levels, endothelin receptor antagonists could be given in parallel to reduce or block any effect of increased endothelin levels. Approaches to increase Aβ catabolism by up-regulation of ECE or other Aβ degrading enzymes such as neprilysin[49] or IDE[50] could be used alone or in conjunction with methods to prevent Aβ production, aggregation, and/or toxicity.

The Clinical Use of ECE Inhibitors Must Be Considered Carefully

ECE inhibitors have received a significant amount of pharmaceutical interest for their potential as anti-hypertensive drugs and for other ailments.[60,68,69] It is likely that if these drugs enter the brain they will increase Aβ levels, potentially leading to the development and/or acceleration of AD in susceptible individuals. Our data argue for a careful evaluation of the effects of these compounds on Aβ accumulation prior to the initiation of human exposure. Based on the genetic mutations that cause AD, in which Aβ levels are elevated for decades prior to the development of the disease, this potential side effect of ECE inhibitors may not be observed for years and must be considered very carefully before clinical use of this class of compounds. The potential risks of neprilysin inhibition must also be carefully considered. Because most ECE and neprilysin inhibitors will inhibit both enzymes at certain concentrations, the use of these drugs may be particularly risky if ECE and neprilysin both play physiological roles in the degradation of Aβ in the brain. ET receptors may be a safer target for pharmacological interference with the endothelin system to reduce hypertension without the side effect of decreased Aβ catabolism.

Acknowledgements

This work was supported by a grant from the Alzheimer's Association to E.E.; Smith Fellowships to C.E. and E.E., a grant from the Bursak Foundation to support the efforts of E.E. and by the Mayo Foundation for Medical Education and Research. We thank Takeda Chemical Industries for their generous gifts of BNT77, BA27, and BC05 and Mona Watson and Cristian-Mihail Prada for their excellent technical assistance.

References

1. Turner AJ, Murphy LJ. Molecular pharmacology of endothelin converting enzymes. Biochem Pharmacol 1996; 51:91-102.
2. Fahnoe DC, Johnson GD, Herman SB et al. Disulfide bonds in big ET-1 are essential for the specific cleavage at the Trp(21)-Val(22) bond by soluble endothelin converting enzyme-1 from baculovirus/insect cells. Arch Biochem Biophys 2000; 373:385-93.
3. Yanagisawa H, Yanagisawa M, Kapur RP et al. Dual genetic pathways of endothelin-mediated intercellular signaling revealed by targeted disruption of endothelin converting enzyme-1 gene. Development 1998; 125:825-36.
4. Hoang MV, Turner AJ. Novel activity of endothelin-converting enzyme: hydrolysis of bradykinin. Biochem J 1997; 327(Pt 1):23-6.
5. Johnson GD, Stevenson T, Ahn K. Hydrolysis of peptide hormones by endothelin-converting enzyme-1. A comparison with neprilysin. J Biol Chem 1999; 274:4053-8.
6. Xu D, Emoto N, Giaid A et al. ECE-1: a membrane-bound metalloprotease that catalyzes the proteolytic activation of big endothelin-1. Cell 1994; 78:473-85.
7. Davenport AP, Kuc RE, Plumpton, C et al. Endothelin-converting enzyme in human tissues. Histochemical Journal 1998; 30:359-74.
8. Korth P, Bohle RM, Corvol P et al. Cellular distribution of endothelin-converting enzyme-1 in human tissues. J Histochem Cytochem 1999; 47:447-62.
9. Valdenaire O, Rohrbacher E, Mattei MG. Organization of the gene encoding the human endothelin-converting enzyme (ECE-1). J Biol Chem 1995; 270:29794-8.
10. Shimada K, Takahashi M, Ikeda M et al. Identification and characterization of two isoforms of an endothelin-converting enzyme-1. FEBS Lett 1995; 371:140-4.
11. Schmidt M, Kroger B, Jacob E et al. Molecular characterization of human and bovine endothelin converting enzyme (ECE-1). FEBS Lett 1994; 356:238-43.
12. Schweizer A, Valdenaire O, Nelbock P et al. Human endothelin-converting enzyme (ECE-1): three isoforms with distinct subcellular localizations. Biochem J 1997; 328(Pt 3):871-7.
13. Valdenaire O, Lepailleur-Enouf D, Egidy G et al. A fourth isoform of endothelin-converting enzyme (ECE-1) is generated from an additional promoter molecular cloning and characterization. Eur J Biochem 1999; 264:341-9.
14. Azarani A, Boileau G, Crine P. Recombinant human endothelin-converting enzyme ECE-1b is located in an intracellular compartment when expressed in polarized Madin-Darby canine kidney cells. Biochem J 1998; 333(Pt 2):439-48.
15. Emoto N, Yanagisawa M. Endothelin-converting enzyme-2 is a membrane-bound, phosphoramidon-sensitive metalloprotease with acidic pH optimum. J Biol Chem 1995; 270:15262-8.
16. Russell FD, Davenport AP. Evidence for intracellular endothelin-converting enzyme-2 expression in cultured human vascular endothelial cells. Circ Res 1999; 84:891-6.
17. Lorenzo MN, Khan RY, Wang Y et al. Human endothelin converting enzyme-2 (ECE2): characterization of mRNA species and chromosomal localization(1). Biochim Biophys Acta 2001; 1522:46-52.
18. Yanagisawa H, Hammer RE, Richardson JA et al. Disruption of ECE-1 and ECE-2 reveals a role for endothelin-converting enzyme-2 in murine cardiac development. J Clin Invest 2000; 105:1373-82.
19. Barnes K, Walkden BJ, Wilkinson TC, Turner AJ. Expression of endothelin-converting enzyme in both neuroblastoma and glial cell lines and its localization in rat hippocampus. J Neurochem 1997; 68:570-7.

20. Sluck JM, Lin RC, Katolik LI et al. Endothelin converting enzyme-1-, endothelin-1-, and endothelin-3-like immunoreactivity in the rat brain. Neuroscience 1999; 91:1483-97.

21. Nakagomi S, Kiryu-Seo S, Kiyama H. Endothelin-converting enzymes and endothelin receptor B messenger RNAs are expressed in different neural cell species and these messenger RNAs are coordinately induced in neurons and astrocytes respectively following nerve injury. Neuroscience 2000; 101:441-9.

22. Hartmann T, Bieger SC, Bruhl B et al. Distinct sites of intracellular production for Alzheimer's disease Aβ40/42 amyloid peptides. Nat Med 1997; 3:1016-20.

23. Yoshizawa T, Iwamoto H, Mizusawa H et al. Cerebrospinal fluid endothelin-1 in Alzheimer's disease and senile dementia of Alzheimer type. Neuropeptides 1992; 22:85-8.

24. Kehoe P, Wavrant-De Vrieze F et al. A full genome scan for late onset Alzheimer's disease. Hum Mol Genet 1999; 8:237-45.

25. Poduslo SE, Yin X, Hargis J et al. A familial case of Alzheimer's disease without tau pathology may be linked with chromosome 3 markers. Hum Genet 1999; 105:32-7.

26. Tanzi RE, Kovacs DM, Kim TW et al. The gene defects responsible for familial Alzheimer's disease. Neurobiol Dis 1996; 3:159-68.

27. Golde TE, Eckman CB, Younkin SG. Biochemical detection of Aβ isoforms: implications for pathogenesis, diagnosis, and treatment of Alzheimer's disease [In Process Citation]. Biochim Biophys Acta 2000; 1502:172-87.

28. Fuller SJ, Storey E, Li QX et al. Intracellular production of beta A4 amyloid of Alzheimer's disease: modulation by phosporamidon and lack of coupling to the secretion of the amyloid precursor protein. Biochemistry 1995; 34:8091-98.

29. Eckman EA, Reed DK, Eckman CB. Degradation of the Alzheimer's amyloid β peptide by endothelin- converting enzyme. J Biol Chem 2001; 276:24540-8.

30. Kukkola PJ, Savage P, Sakane Y. Differential structure-activity relationships of phosphoramidon analogues for inhibition of three metalloproteases: endothelin- converting enzyme, neutral endopeptidase, and angiotensin-converting enzyme. J Cardiovasc Pharmacol 1995; 26(Suppl 3):S65-8.

31. Ansorge S, Bohley P, Kirschke H et al. The insulin and glucagon degrading proteinase of rat liver: a metal-dependent enzyme. Biomed Biochim Acta 1984; 43:39-46.

32. Howell S, Nalbantoglu J, Crine P. Neutral endopeptidase can hydrolyze beta-amyloid(1-40) but shows no effect on beta-amyloid precursor protein metabolism. Peptides 1995; 16:647-52.

33. Iwata N, Tsubuki S, Takaki Y et al. Identification of the major Aβ$_{1-42}$-degrading catabolic pathway in brain parenchyma: suppression leads to biochemical and pathological deposition. Nat Med 2000; 6:143-50.

34. Hu J, Igarashi A, Kamata M et al. Angiotensin-converting enzyme degrades Alzheimer amyloid β-peptide Aβ, retards Aβ aggregation, deposition, fibril formation and inhibits cytotoxicity. J Biol Chem 2001; 16:16.

35. Turner AJ, Tanzawa K. Mammalian membrane metallopeptidases: NEP, ECE, KELL, and PEX. Faseb J 1997; 11:355-64.

36. Corvol P, Williams TA, Soubrier F. Peptidyl dipeptidase A: angiotensin I-converting enzyme. Methods Enzymol 1995; 248:283-305.

37. Spillantini MG, Sicuteri F, Salmon S, et al. Characterization of endopeptidase 3.4.24.11 ("enkephalinase") activity in human plasma and cerebrospinal fluid. Biochem Pharmacol 1990; 39:1353-6.

38. Soleilhac JM, Lafuma C, Porcher JM, et al. Characterization of a soluble form of neutral endopeptidase-24.11 (EC 3.4.24.11) in human serum: enhancement of its activity in serum of underground miners exposed to coal dust particles. Eur J Clin Invest 1996; 26:1011-7.

39. Ikeda K, Emoto N, Raharjo SB et al. Molecular identification and characterization of novel membrane-bound metalloprotease, the soluble secreted form of which hydrolyzes a variety of vasoactive peptides. J Biol Chem 1999; 274:32469-77.

40. Ghaddar G, Ruchon AF, Carpentier M et al. Molecular cloning and biochemical characterization of a new mouse testis soluble-zinc-metallopeptidase of the neprilysin family. Biochem J 2000; 347(Pt 2):419-29.

41. Tanja O, Facchinetti P, Rose C et al. Neprilysin II: A putative novel metalloprotease and its isoforms in CNS and testis. Biochem Biophys Res Commun 2000; 271:565-70.

42. Roques BP, Beaumont A. Neutral endopeptidase-24.11 inhibitors: from analgesics to antihypertensives? Trends Pharmacol Sci 1990; 11:245-9.
43. Gronhagen-Riska C, Fyhrquist F. Purification of human lung angiotensin-converting enzyme. Scand J Clin Lab Invest 1980; 40:711-9.
44. Ahn K, Herman SB, Fahnoe DC. Soluble human endothelin-converting enzyme-1: expression, purification, and demonstration of pronounced pH sensitivity. Arch Biochem Biophys 1998; 359:258-68.
45. Ahn K, Pan S, Beningo K et al. A permanent human cell line (EA.hy926) preserves the characteristics of endothelin converting enzyme from primary human umbilical vein endothelial cells. Life Sci 1995; 56:2331-41.
46. Parnot C, Le Moullec JM, Cousin MA et al. A live-cell assay for studying extracellular and intracellular endothelin-converting enzyme activity. Hypertension 1997; 30:837-44.
47. Korth P, Egidy G, Parnot C, et al. Construction, expression and characterization of a soluble form of human endothelin-converting-enzyme-1. FEBS Lett 1997; 417:365-70.
48. Fahnoe DC, Knapp J, Johnson GD et al. Inhibitor potencies and substrate preference for endothelin-converting enzyme-1 are dramatically affected by pH. J Cardiovasc Pharmacol 2000; 36(5 Suppl 1):S22-5.
49. Iwata N, Tsubuki S, Takaki Y et al. Metabolic regulation of brain Abeta by neprilysin. Science 2001; 292:1550-2.
50. Selkoe DJ. Clearing the Brain's Amyloid Cobwebs. Neuron 2001; 32:177-80.
51. Orzechowski HD, Gunther A, Menzel S et al. Endothelial expression of endothelin-converting enzyme-1 beta mRNA is regulated by the transcription factor Ets-1. J Cardiovasc Pharmacol 1998; 31(Suppl 1):S55-7.
52. Orzechowski HD, Richter CM, Funke-Kaiser H et al. Evidence of alternative promoters directing isoform-specific expression of human endothelin-converting enzyme-1 mRNA in cultured endothelial cells. J Mol Med 1997; 75:512-21.
53. Funke-Kaiser H, Orzechowski HD, Richter M et al. Human endothelin-converting enzyme-1 β mRNA expression is regulated by an alternative promoter. J Cardiovasc Pharmacol 1998; 31(Suppl 1):S7-9.
54. Barker S, Khan NQ, Wood EG et al. Effect of an antisense oligodeoxynucleotide to endothelin-converting enzyme-1c (ECE-1c) on ECE-1c mRNA, ECE-1 protein and endothelin-1 synthesis in bovine pulmonary artery smooth muscle cells. Mol Pharmacol 2001; 59:163-9.
55. Shao R, Yan W, Rockey DC. Regulation of endothelin-1 synthesis by endothelin-converting enzyme-1 during wound healing. J Biol Chem 1999; 274:3228-34.
56. Schweizer A, Loffler BM, Rohrer J. Palmitoylation of the three isoforms of human endothelin-converting enzyme-1. Biochem J 1999; 340(Pt 3):649-56.
57. Nelboeck P, Fuchs M, Bur D et al. Glycosylation of Asn-632 and Asn-651 is important for functional expression of endothelin-converting enzyme-1. J Cardiovasc Pharmacol 1998; 31(Suppl 1):S4-6.
58. Ehrenreich H, Loffler BM, Hasselblatt M et al. Endothelin converting enzyme activity in primary rat astrocytes is modulated by endothelin B receptors. Biochem Biophys Res Commun 1999; 261:149-55.
59. Lopez-Ongil S, Senchak V, Saura M et al. Superoxide regulation of endothelin-converting enzyme. J Biol Chem 2000; 275:26423-7.
60. Loffler BM. Endothelin-converting enzyme inhibitors: current status and perspectives. J Cardiovasc Pharmacol 2000; 35:S79-82.
61. Nishikawa S, Miyamoto A, Yamamoto H et al. Preeclamptic serum enhances endothelin-converting enzyme expression in cultured endothelial cells. Am J Hypertens 2001; 14:77-83.
62. Kwan AL, Bavbek M, Jeng AY et al. Prevention and reversal of cerebral vasospasm by an endothelin-converting enzyme inhibitor, CGS 26303, in an experimental model of subarachnoid hemorrhage. J Neurosurg 1997; 87:281-6.
63. Eto M, Barandier C, Rathgeb L et al. Thrombin suppresses endothelial nitric oxide synthase and upregulates endothelin-converting enzyme-1 expression by distinct pathways: role of Rho/ROCK and mitogen-activated protein kinase. Circ Res 2001; 89:583-90.
64. Naomi S, Iwaoka T, Disashi T et al. Endothelin-1 inhibits endothelin-converting enzyme-1 expression in cultured rat pulmonary endothelial cells. Circulation 1998; 97:234-6.

65. Barton M, Shaw S, d'Uscio LV et al. Angiotensin II increases vascular and renal endothelin-1 and functional endothelin converting enzyme activity in vivo: role of ETA receptors for endothelin regulation. Biochem Biophys Res Commun 1997; 238:861-5.

66. Morawietz H, Goettsch W, Szibor M et al. Angiotensin-converting enzyme inhibitor therapy prevents upregulation of endothelin-converting enzyme-1 in failing human myocardium. Biochem Biophys Res Commun 2002; 295:1057-61.

67. Telemaque S, Emoto N, deWit D et al. In vivo role of endothelin-converting enzyme-1 as examined by adenovirus-mediated overexpression in rats. J Cardiovasc Pharmacol 1998; 31(Suppl 1):S548-50.

68. Gray GA, Webb DJ. The endothelin system and its potential as a therapeutic target in cardiovascular disease. Pharmacol Ther 1996; 72:109-48.

69. Kitas EA, Loffler BM, Daetwyler S et al. Synthesis of triazole-Tethered pyrrolidine libraries: novel ECE inhibitors. Bioorg Med Chem Lett 2002; 12:1727-30.

CHAPTER 8

Aβ Metabolism in Cholesterol Enriched Membrane Microdomains

Todd E. Golde and M. Paul Murphy

Abstract

Aggregation and accumulation of Aβ is a pathological hallmark of the Alzheimer's disease (AD) brain. Studies on cerebrospinal fluid (CSF) from living patients and postmortem control brain tissue show the normal level of Aβ in the brain to be in the low nM range. Biophysical studies of Aβ aggregation indicate that in the absence of other factors monomeric Aβ does not readily aggregate into amyloid fibrils unless it is present at μM concentrations. Based on these observations it seems that other factors must exist within the brain that either result in higher local concentrations of Aβ, catalyze Aβ fibril formation, or both. One possibility is that Aβ fibril formation takes place in buoyant cholesterol enriched membrane microdomains (CEMs). There is evidence that Aβ is produced in these domains, and that factors that foster Aβ aggregation such as cholesterol and GM1 ganglioside are concentrated in these domains. Further studies of Aβ production, degradation, and clearance in CEMs may reveal important clues as to how Aβ aggregation occurs. These studies could lead to novel therapeutic approaches for the treatment of AD.

Introduction

There is compelling evidence that abnormal accumulation of the amyloid β protein (Aβ) plays a causal role in the development of Alzheimer's disease (AD) (reviewed in ref. 1). Because of the presumed causal role of Aβ accumulation in AD, a number of therapies targeting Aβ production, aggregation, or clearance are being considered. Despite impressive advances in the preclinical development of therapeutic strategies targeting Aβ, to date no anti-Aβ therapy has been successfully tested in the clinic. Moreover, despite the enormous advances in our understanding of Aβ metabolism that have been made over the last several years, many questions regarding the normal metabolism of Aβ and its pathologic metabolism in AD remain unanswered. One of the central questions in this regard is how does Aβ aggregate in vivo. Average concentrations of Aβ in the brain and cerebrospinal fluid (CSF) do not appear sufficient to promote Aβ aggregation into amyloid.[2-9] Multiple factors within the brain exist that could promote the initial steps of Aβ aggregation. Surprisingly, there is evidence that many of these Aβ promoting factors exist within buoyant cholesterol rich membranes. Thus, we postulate that factors within these membranes may promote the formation of Aβ aggregates, possibly the most critical step in AD pathogenesis.

Aβ Metabolism and Alzheimer's Disease, edited by Takaomi C. Saido. ©2003 Eurekah.com.

CEMs as a Generic Term for Cholesterol Enriched Membrane Microdomains

For those unfamiliar with membrane microdomains, the associated terminology can be quite confusing. We propose the term cholesterol enriched membrane microdomains (CEMs) as a general term applicable to any cholesterol and sphingolipid enriched membrane microdomain. Over the years, various terms have been used to describe CEMs, including caveolar membranes, detergent resistant membranes, glycosphingolipid enriched membranes and lipid rafts (reviewed in ref. 10). These descriptive terms stem from the original characterization of these various types of membrane microdomains, and although each has been used to describe a specific type of membrane microdomain, they are often used with less specificity than originally intended.

Some of these terms refer to the biochemical properties of these microdomains. For example, the term "detergent resistant membrane" refers to the fact that these membranes can be isolated after detergent homogenization by flotation to a low-density region of a sucrose density gradient by ultracentrifugation.[10] This technique is often used as a first step towards defining proteins that may be associated with these membranes. Other terms such as caveolar membranes are descriptive terms for a certain type of CEM. Caveolae are flask shaped invaginations found on the plasma membrane. They are enriched for the protein caveolin (or in the case of the brain, flotillin) and are thought to play a role in nonclathrin coat mediated endo- and transcytotic events.[10] Lipid rafts are a more functional description that refers to dynamic assemblies of membranes enriched in cholesterol and sphingolipids that cluster proteins such as glycosylphosphatidylinositol anchored proteins and signaling molecules in the membrane.[10] Although CEMs and lipid rafts might be considered interchangeable terms, it is not clear that all buoyant cholesterol enriched membranes are rafts, as a rather specific definition of rafts and subsets of rafts has been proposed with an emphasis on functional as well as structural properties.[10]

The CEM or raft concept has been controversial. It had previously been argued that the apparent properties of the microdomains were experimental artifacts and not present in living cells. However, recent studies reviewed in Simons and Toomre have largely dispelled most of these concerns.[10] Briefly, these studies demonstrated that proteins cluster in CEMs in cells and that certain proteins are found in CEMs and others are not. CEMs are dynamic, not static, structures that can form rapidly and then dissipate. With respect to the study of proteins involved in AD, most of the studies of CEMs can be viewed as incomplete, as they have relied on more traditional techniques such as isolation of detergent resistant membranes to demonstrate association with CEMs.

One key concept regarding CEMs is that they are functionally important. CEMs have been implicated in mediating a range of signaling events and various aspects of organelle trafficking.[10] It is thought that through the clustering of important proteins these microdomains favor specific protein-protein interactions that result in efficient cell signaling. As described below, this clustering may also facilitate the proteolytic processing of various membrane bound proteins. Insight into CEM function can be gained by disrupting the membranous components of the CEM. Typical methods to disrupt CEMs involve cholesterol sequestration, depletion, or inhibition of cholesterol synthesis. Other methods less commonly used involve alteration of other CEM components, such as gangliosides.

Aβ Production in CEMs

Two proteases produce Aβ from the amyloid β protein precursor (APP) through sequential cleavages (reviewed in ref.11). APP is first cleaved by β-secretase (BACE1, Asp-2, memapsin-1), a transmembrane aspartyl protease, at the amino terminus of Aβ to generate a large secreted derivative (sAPPβ) and a membrane bound APP carboxyl terminal fragment

(CTFβ). Subsequent cleavage of CTFβ by γ-secretase results in production of the Aβ peptide and CTFγ. In a second pathway, APP is cleaved within the Aβ sequence by α-secretase, which generates another large secreted derivative and CTF (sAPPα and CTFα).

Recent evidence indicates that the first cleavage step in Aβ generation (Fig. 1), β-secretase cleavage, may occur in CEMs. β-secretase is enriched in CEMs that are distinct from caveolar containing CEMs.[12] Although β-secretase activity was not measured, the concentration of mature β-secretase in these membranes provides initial evidence that this cleavage may occur at this site. This localization would also be consistent with the observation that lowering cholesterol reduces β-secretase cleavage, described in detail below. In addition, there is evidence that alterations in caveolin-3 expression can alter β-secretase cleavage of APP.[13] How this relates to the presence of β-secretase in noncaveolar CEMs is not clear.

γ-secretase activity is also enriched in CEMs (Fig. 1), as are two proteins necessary for γ-secretase activity, presenilins (PS) and nicastrin.[14,15] PS are polytopic membrane proteins that appear to be the catalytic subunits of a high molecular weight complex (reviewed in ref. 11). Although the precise catalytic mechanism of γ-secretase is not yet established, there is evidence that supports the concept that PS and γ-secretase represent a novel type of aspartyl protease. Nicastrin is predicted to be a type 1 transmembrane glycoprotein and its expression is essential for γ-secretase activity and PS complex formation.[16-18] Its mechanistic role in γ-secretase cleavage is not known. In human and mouse brain, almost all of the nicastrin is detected in CEMs that possess γ-secretase activity (Murphy and Golde, unpublished observation). Upon overexpression in cultured cells wild-type, but not mutant, nicastrin increases Aβ production. In these cell culture experiments, mature forms of wild-type nicastrin is present in CEMs whereas mutant nicastrin is not. Localization of γ-secretase in the CEMs also appears to be of critical functional importance. Depletion of cholesterol from CEMs results in almost complete reduction in γ-secretase activity, and replenishment of cholesterol restores activity.[14]

Although previous studies of CEMs have focused on domains that are found in the plasma membrane, Golgi, or vesicles of the secretory and endocytic pathway, these studies on PS, Nicastrin and γ-secretase activity indicate that CEM domains must be present in the endoplasmic reticulum (ER), as a majority of PS is found in the ER.[14,15] Importantly, immunoelectron microscopic studies of PS1 in cultured cells reveals a patched distribution of PS1 in the ER and ER Golgi trafficking intermediates, suggesting that PS1 is normally clustered in membrane microdomains.[19] However, these studies do not prove that the ER localized PS is responsible for γ-secretase activity. It is more likely that a small amount of PS within the γ-secretase complex in other organelles is responsible for the majority of γ-secretase cleavage of APP.

Studies in cell culture have demonstrated that alterations in cholesterol levels can alter Aβ production (reviewed in ref. 20). Depletion of cellular cholesterol decreases Aβ production and can, under certain circumstances, increase nonamyloidogenic processing by α-secretase.[21-23] Although depletion of cellular cholesterol in vivo does appear to decrease γ-activity, the major effect of this reduction is to decrease β-secretase cleavage of APP. More recently, it has been shown that treatment of animals with HMG-Co-A reductase inhibitors (also known as statins) can decrease Aβ production by reducing amyloidogenic β-secretase processing.[24] In the statin treated animals Aβ levels decreased in parallel with cholesterol.

Because cholesterol is an essential component of CEMs and since the removal of cholesterol from these membranes disrupts their function, it is possible that the effect of cholesterol on secretase activity can largely be explained by APP processing events that are associated with CEMs. Additional studies will be needed to determine if physiologic alterations in cholesterol levels can influence cholesterol levels in CEMs. To date, there have been no studies exploring whether modulation of cholesterol levels in vivo can alter CEMs and their composition. Nevertheless, there is an intriguing correlation between amyloidogenic processing of Aβ in CEMs and alterations in Aβ production that occur when cholesterol levels are altered.

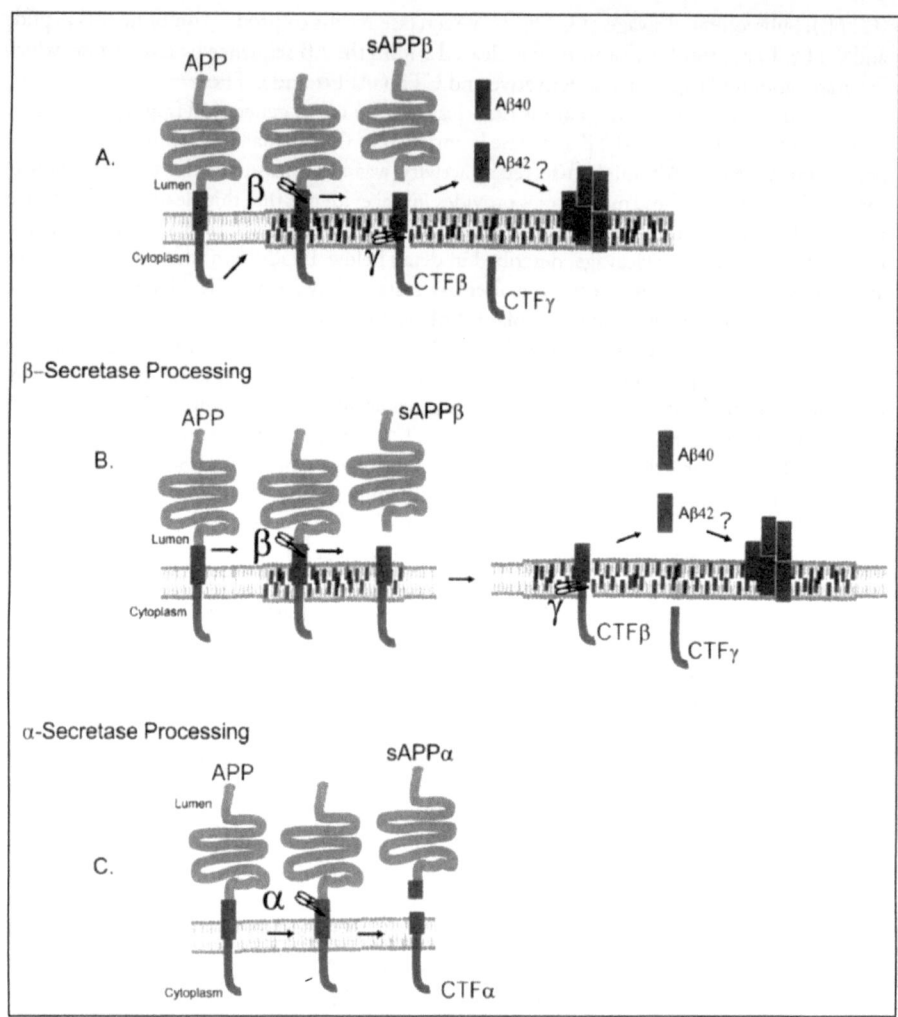

Figure 8.1. APP processing in CEMs. The amyloid protein precursor (APP) is a type I transmembrane protein that is processed in several different pathways. Generation of the amyloid β protein (Aβ) in the β-secretase pathway (A and B) requires two proteolytic events, a proteolytic cleavage at the amino terminus of the Aβ sequence, referred to as β-secretase cleavage and a cleavage at the carboxyl terminus, known as γ-secretase cleavage. Cleavage by β-secretase results in the secretion of sAPPβ and production of the membrane bound carboxyl terminal fragment β (CTFβ). γ-secretase cleavage of CTFβ produces the secreted Aβ peptide and the CTFγ. In the α-secretase pathway (C), the APP is cleaved within Aβ to generate a large, secreted derivative referred to as sAPPα and a membrane associated CTFα. Aβ production in the β-secretase pathway appears to occur in CEMs that are indicated by the presence of high levels of cholesterol in the membrane (light gray squares) and GM1 ganglioside (dark gray squares). It is not certain whether the CEMs that contain β- and γ-secretase activity are contiguous (A) or spatially distinct (B). Local production of Aβ in CEMs (A or B) could result in local aggregation due to the high concentrations of Aβ and the fibril promoting factors present in CEMs. In non-CEM membranes, the α-secretase pathway is favored (C).

Estimates of Aβ levels in human control brain suggest that it is present at low nanomolar levels.[2,3] This level of Aβ can drive the aggregation of Aβ into soluble globular Aβ oligomers that are also referred to as ADDLs.[25] However, based on in vitro aggregation studies, such concentrations are not sufficient to foster deposition of Aβ as insoluble fibrillar amyloid aggregates. Local production of Aβ in CEMs, especially if it occurs in a secretory vesicle, may produce the locally high concentrations of Aβ that could support fibril formation.

Aβ Deposition, Aggregation and CEMs

Although production of Aβ in CEMs could produce high enough local concentrations to foster deposition, it is likely that other factors promote Aβ aggregation. Two factors enriched in brain CEMs that can be shown to promote Aβ aggregation in vitro are cholesterol and GM-1 ganglioside.[15,26-33] Aβ can be shown to accumulate in CEMs in the AD brain and in transgenic mouse models of AD that develop age dependent Aβ deposits. In AD brain and AD transgenic mouse models, some of the Aβ that accumulates in CEMs appears to be tightly associated with GM-1 ganglioside.[31] Although not studied in detail, GM-1 levels are high in regions of the brain prone to Aβ deposition, such as the hippocampus and cerebral cortex, whereas regions relatively resistant to Aβ deposition, such as the cerebellum, have lower levels.[34]

Indirect evidence suggests that alterations in cholesterol can influence Aβ deposition in vivo. Increasing cholesterol increases amyloid deposition in transgenic mice[35] and decreasing cholesterol decreases Aβ deposition and Aβ production.[36] Accumulation of cholesterol in Neimann Pick type C mice is also associated with Aβ accumulation.[32] In these mice, the Aβ accumulates along with cholesterol in CEMs that appear to reside in endocytic organelles.

ApoE, Cholesterol and CEMs

Genetic studies of late-onset AD provided one of the first clues that cholesterol might play a role in AD pathology, when it was discovered that the E4 isoform of apolipoprotein E (ApoE) is a risk factor for AD.[37] ApoE is a plasma lipoprotein, but is also the major lipoprotein expressed in the brain. Outside the brain, ApoE plays a basic role in the degradation of particles rich in cholesterol and triglycerides.[38] It is able to bind to LDL receptors and receptors for chylomicron remnants. There are three major ApoE isoforms, E2, E3, and E4. The E3 allele is the predominant form and E3 affects metabolism of lipoproteins in a standard way. When compared to the E3 allele, the E2 allele is associated with lower LDL levels, whereas E4 is associated with higher LDL levels. This has some impact on the progression of arteriosclerosis and is probably responsible for the modest increase in risk for cardiac events in patients with the E4 allele, and the slight protective effect of the E2 allele.[38] Similarly, ApoE2 appears to be associated with a decreased risk for AD.[39]

Although the associations between the risk for atherosclerotic disease, AD and ApoE are intriguing, it is not certain that the increased risk for AD associated with the ApoE4 allele is attributable to alterations in cholesterol. If so, then the very small differences in plasma cholesterol associated with ApoE4 can increase the risk for AD. Furthermore, there is no evidence available that would support the notion that the risk for AD could be attributable to changes in brain cholesterol. ApoE knockout mice show no changes in total brain cholesterol levels, despite the fact that plasma cholesterol is elevated, arguing that ApoE genotype is unlikely to markedly influence total brain cholesterol, although it is possible that more subtle alterations in cholesterol metabolism exist, such as alterations in sterol recycling (ref. 40 and D. Holtzman, personal communication). Studies on transgenic and knockout mice have revealed that ApoE is a necessary factor for fibrillar amyloid deposition, since in its absence Aβ deposits only in nonfibrillar forms.[41] When human ApoE3 or ApoE4 are expressed under control of brain specific promoters in an ApoE knockout background, and these mice are then crossed to mice overexpressing APP, fibrillar Aβ deposits do form.[40,42] Moreover, in mice expressing ApoE4, these Aβ fibrils form faster than in mice expressing ApoE3.[40,42-44] Together with in vitro studies suggesting that ApoE might promote Aβ fibrillization, these studies suggest that ApoE plays a direct role in Aβ aggregation.[45] Thus, although it has not been excluded that the increased risk for AD associated with ApoE4 is attributable to alterations in cholesterol levels, it appears that the effect of ApoE4 is to directly alter Aβ aggregation.

With respect to ApoE and CEMs, there is some initial data to suggest that ApoE could influence the Aβ accumulation in raft-like domains. Alterations in ApoE expression do appear to alter Aβ deposition within CEMs in APP transgenic mice.[40] However, at present no mechanistic insight into this potentially interesting phenomenon has been provided.

Aβ Degrading Enzymes in CEMs

Aβ accumulation occurs when there is an imbalance between production and normal clearance mechanisms. Multiple proteases have been implicated in Aβ clearance including but not limited to neprilysin, insulin degrading enzyme, endothelin converting enzyme, and plasmin. [46-55] CEMs have been shown to cluster many peptidases,[56] and there is experimental data showing that both plasmin and neprilysin are localized to CEMs.[49,57] Although enzymes for both production and degradation of Aβ exist in CEMs, it is not known whether production and degradation occurs in the same CEM or spatially distinct CEMs. Future studies will be needed to address this issue.

Summary

Many of the proteins and co-factors that play a central role in Aβ metabolism appear to be either enriched in or exclusively present in CEMs. Aβ appears to be produced in these microdomains as both β- and γ-secretase activities appear to localize to these domains. Although the functional consequences of this localization are not well understood, it appears that the increased level of cholesterol in these membranes is important for proper γ-secretase cleavage of APP. Perhaps CEM integrity and lipid composition create an optimal environment for amyloidogenic processing of APP by clustering proteases and substrates in an optimal configuration. Thus, promoting CEM integrity or number, by increasing cholesterol or other factors associated with these domains, would increase amyloidogenic cleavage, whereas disrupting CEMs would inhibit cleavage and in some circumstances lead to enhanced nonamyloidogenic cleavage by α-secretase. On the other hand, there is some evidence that certain Aβ degrading enzymes also cluster in these rafts. Could this clustering be protective? If locally high Aβ concentrations are made at a site that would promote aggregation, then it certainly would seem appropriate to try to prevent this aggregation by rapidly clearing Aβ.

Given the epidemiologic data showing that increased cholesterol levels are a risk factor for AD and studies that suggest that lowering cholesterol may be protective,[58-60] there would seem to be a link between Aβ metabolism in CEMs, cholesterol and factors such as ApoE that influence cholesterol metabolism. However, additional studies will be needed to determine if physiologic alterations of in vivo cholesterol levels can influence APP distribution within raft domains. To date, there have been no studies exploring whether modulation of cholesterol levels in vivo can alter raft function and composition.

Understanding the factors that foster Aβ accumulation in the brain is likely to play a critical role in our ability to develop AD therapies aimed at preventing Aβ accumulation. The emerging data reviewed herein suggest that understanding how Aβ is metabolized in CEMs may provide critical insights into this process. Furthermore, if Aβ accumulation occurs in CEMs in the AD brain, then it is possible that this accumulation will have functional consequences. Although there is no data to support this speculation, given the role of CEMs in mediating critical cell signaling events, the functional consequences of Aβ accumulation in these domains should be examined.

References

1. Selkoe DJ. Alzheimer's disease: genes, proteins, and therapy. Physiol Rev 2001; 81:741-66.
2. Seubert P, Vigo-Pelfrey C, Esch F et al. Isolation and quantitation of soluble Alzheimer's β-peptide from biological fluids. Nature 1992; 359:325-27.
3. Gravina SA, Ho L, Eckman CB et al. Amyloid beta protein (Aβ) in Alzheimer's disease brain. Biochemical and immunocytochemical analysis with antibodies specific for forms ending at $A\beta_{40}$ or $A\beta_{42(43)}$. Journal of Biological Chemistry 1995; 270:7013-6.
4. Hosoda R, Saido TC, Otvos L, Jr et al. Quantification of modified amyloid β peptides in Alzheimer disease and Down syndrome brains. Journal of Neuropathology & Experimental Neurology 1998; 57:1089-95.
5. Haass C, Schlossmacher MG, Hung AY et al. Amyloid β-peptide is produced by cultured cells during normal metabolism. Nature 1992; 359:322-25.
6. Shoji M, Golde TE, Ghiso J et al. Production of the Alzheimer amyloid β protein by normal proteolytic processing. Science 1992; 258:126-9.
7. Jarrett JT, Berger EP Jr PTL. The carboxy terminus of β amyloid protein is critical for the seeding of amyloid formation: Implications for pathogenesis of Alzheimer's disease. Biochem 1993; 32:4693-97.
8. Evans KC, Berger EP, Cho CG et al. Apolipoprotein E is a kinetic but not a thermodynamic inhibitor of amyloid formation: implications for the pathogenesis and treatment of Alzheimer disease. Proc Natl Acad Sci USA 1995; 92:763-7.
9. Kirkitadze MD, Condron MM, Teplow DB. Identification and characterization of key kinetic intermediates in amyloid β-protein fibrillogenesis. J Mol Biol 2001; 312:1103-19.
10. Simons K, Toomre D. Lipid rafts and Signal Transduction. Nat Rev Mol Cell Biol 2000; 1:31-39.
11. Golde TE, Younkin SG. Presenilins as therapeutic targets for the treatment of Alzheimer's disease. Trends Mol Med 2001; 7:264-9.
12. Riddell DR, Christie G, Hussain I et al. Compartmentalization of β-secretase (Asp2) into low-buoyant density, noncaveolar lipid rafts. Curr Biol 2001; 11:1288-93.
13. Nishiyama K, Trapp BD, Ikezu T et al. Caveolin-3 upregulation activates beta-secretase-mediated cleavage of the amyloid precursor protein in Alzheimer's disease. J Neurosci 1999; 19:6538-48.
14. Wahrle S, Das P, Nyborg AC et al. Cholesterol-Dependent gamma-Secretase Activity in Buoyant Cholesterol- Rich Membrane Microdomains. Neurobiol Dis 2002; 9:11-23.
15. Lee SJ, Liyanage U, Bickel PE et al. A detergent-insoluble membrane compartment contains Aβ in vivo. Nature Medicine 1998; 4:730-4.
16. Yu G, Nishimura M, Arawaka S et al. Nicastrin modulates presenilin-mediated notch/glp-1 signal transduction and βAPP processing [see comments]. Nature 2000; 407:48-54.
17. Kaether C, Lammich S, Edbauer D et al. Presenilin-1 affects trafficking and processing of betaAPP and is targeted in a complex with nicastrin to the plasma membrane. J Cell Biol 2002; 158:551-61.
18. Edbauer D, Winkler E, Haass C et al. Presenilin and nicastrin regulate each other and determine amyloid β-peptide production via complex formation. Proc Natl Acad Sci USA 2002; 99:8666-71.
19. Kim SH, Lah JJ, Thinakaran G et al. Subcellular localization of presenilins: association with a unique membrane pool in cultured cells. Neurobiol Dis 2000; 7:99-117.
20. Golde TE, Eckman CB. Cholesterol modulation as an emerging strategy for the treatment of Alzheimer's disease. Drug Discov Today 2001; 6:1049-55.
21. Simons M, Keller P, De Strooper B et al. Cholesterol depletion inhibits the generation of β-amyloid in hippocampal neurons. Proc Natl Acad Sci USA 1998; 95:6460-4.
22. Kojro E, Gimpl G, Lammich S et al. Low cholesterol stimulates the nonamyloidogenic pathway by its effect on the α-secretase ADAM 10. Proc Natl Acad Sci USA 2001; 98:5815-20.
23. Bodovitz S, Klein WL. Cholesterol modulates a-secretase cleavage of amyloid precursor protein. Journal of Biological Chemistry 1996; 271:4436-40.
24. Fassbender K, Simons M, Bergmann C et al. Simvastatin strongly reduces levels of Alzheimer's disease β-amyloid peptides $A\beta_{42}$ and $A\beta_{40}$ in vitro and in vivo. Proc Natl Acad Sci USA 2001; 98:5856-61.
25. Lambert MP, Barlow AK, Chromy BA, Edwards C, Freed R, Liosatos M et al. Diffusible, nonfibrillar ligands derived from $A\beta_{1-42}$ are potent central nervous system neurotoxins. Proc Natl Acad Sci USA 1998; 95:6448-53.

26. Choo-Smith LP, Surewicz WK. The interaction between Alzheimer amyloid β $_{(1-40)}$ peptide and ganglioside GM1-containing membranes. FEBS Letters 1997; 402:95-8.

27. Choo-Smith LP, Garzon-Rodriguez W, Glabe CG et al. Acceleration of amyloid fibril formation by specific binding of Aβ $_{(1-40)}$ peptide to ganglioside-containing membrane vesicles. J Biol Chem 1997;272:22987-90.

28. Avdulov NA, Chochina SV, Igbavboa U et al. Lipid binding to amyloid β-peptide aggregates: preferential binding of cholesterol as compared with phosphatidylcholine and fatty acids. J Neurochem 1997; 69:1746-52.

29. Yanagisawa K, Ihara Y. GM1 ganglioside-bound amyloid β-protein in Alzheimer's disease brain. Neurobiology of Aging 1998; 19(1 Suppl):S65-7.

30. Yanagisawa K, McLaurin J, Michikawa M et al. Amyloid β -protein (Aβ) associated with lipid molecules: immunoreactivity distinct from that of soluble Ab. FEBS Letters 1997; 420:43-6.

31. Yanagisawa K, Odaka A, Suzuki N, Ihara Y. GM1 ganglioside-bound amyloid beta-protein (A beta): a possible form of preamyloid in Alzheimer's disease [see comments]. Nature Medicine 1995; 1:1062-6.

32. Yamazaki T, Chang TY, Haass C et al. Accumulation and aggregation of amyloid β-protein in late endosomes of Niemann-pick type C cells. J Biol Chem 2001; 276:4454-60.

33. Kakio A, Nishimoto Si S, Yanagisawa K et al. Cholesterol-dependent Formation of GM1 Ganglioside-bound Amyloid β-Protein, an Endogenous Seed for Alzheimer Amyloid. J Biol Chem 2001; 276:24985-90.

34. Kracun I, Rosner H, Cosovic C et al. Topographical atlas of the gangliosides of the adult human brain. J Neurochem 1984; 43:979-89.

35. Refolo LM, Pappolla MA, Malester B et al. Hypercholesterolemia accelerates the Alzheimer's amyloid pathology in a transgenic mouse model. Neurobiol Dis 2000; 7:321-31.

36. Refolo LM, Pappolla MA, LaFrancois J et al. A cholesterol-lowering drug reduces β-amyloid pathology in a transgenic mouse model of Alzheimer's disease. Neurobiol Dis 2001; 8:890-9.

37. Strittmatter WJ, Roses AD. Apolipoprotein E and Alzheimer disease. [Review]. Proceedings of the National Academy of Sciences of the United States of America 1995; 92:4725-7.

38. Breslow JL. Genetics of lipoprotein abnormalities associated with coronary artery disease susceptibility. Annu Rev Genet 2000; 34:233-54.

39. Roses AD. Apolipoprotein E, a gene with complex biological interactions in the aging brain. Neurobiology of Disease 1997; 4:170-85.

40. Fagan AM, Watson M, Parsadanian M et al. Human and murine ApoE markedly alters Aβ metabolism before and after plaque formation in a mouse model of Alzheimer's disease. Neurobiol Dis 2002; 9:305-18.

41. Bales K, Verina T, Dodel R et al. Lack of apolipoprotein E dramatically reduces amyloid β-peptide deposition. Nature Genetics 1997; 17:263-64.

42. Holtzman DM, Bales KR, Wu S et al. Expression of human apolipoprotein E reduces amyloid β deposition in a mouse model of Alzheimer's disease. J Clin Invest 1999; 103:R15-R21.

43. Holtzman DM, Fagan AM, Mackey B et al. Apolipoprotein E facilitates neuritic and cerebrovascular plaque formation in an Alzheimer's disease model. Ann Neurol 2000; 47:739-47.

44. Holtzman DM, Bales KR, Tenkova T et al. Apolipoprotein E isoform-dependent amyloid deposition and neuritic degeneration in a mouse model of Alzheimer's disease. Proc Natl Acad Sci USA 2000; 97:2892-7.

45. Castano EM, Prelli F, Wisniewski T et al. Fibrillogenesis in Alzheimer's disease of amyloid β peptides and apolipoprotein E. Biochemical Journal 1995; 306:599-604.

46. Selkoe DJ. Clearing the brain's amyloid cobwebs. Neuron 2001; 32:177-80.

47. Iwata N, Tsubuki S, Takaki Y et al. Identification of the major Aβ$_{1-42}$-degrading catabolic pathway in brain parenchyma: suppression leads to biochemical and pathological deposition. Nat Med 2000; 6:143-50.

48. Fukami S, Watanabe K, Iwata N et al. Aβ-degrading endopeptidase, neprilysin, in mouse brain: synaptic and axonal localization inversely correlating with Aβ pathology. Neurosci Res 2002; 43:39-56.

49. Shirotani K, Tsubuki S, Iwata N et al. Neprilysin degrades both amyloid β peptides 1-40 and 1-42 most rapidly and efficiently among thiorphan- and phosphoramidon-sensitive endopeptidases. J Biol Chem 2001; 276:21895-901.

50. Eckman EA, Reed DK, Eckman CB. Degradation of the Alzheimer's amyloid β peptide by endothelin- converting enzyme. J Biol Chem 2001; 276:24540-8.

51. Chesneau V, Vekrellis K, Rosner MR et al. Purified recombinant insulin-degrading enzyme degrades amyloid β protein but does not promote its oligomerization. Biochem J 2000; 351 Pt 2:509-16.

52. Vekrellis K, Ye Z, Qiu WQ et al. Neurons regulate extracellular levels of amyloid β protein via proteolysis by insulin-degrading enzyme. J Neurosci 2000; 20:1657-65.

53. Qiu WQ, Walsh DM, Ye Z et al. Insulin-degrading enzyme regulates extracellular levels of amyloid β- protein by degradation. J Biol Chem 1998; 273:32730-8.

54. Tucker HM, Kihiko-Ehmann M, Wright S et al. Tissue plasminogen activator requires plasminogen to modulate amyloid- beta neurotoxicity and deposition. J Neurochem 2000; 75:2172-7.

55. Tucker HM, Kihiko M, Caldwell JN, Wright S et al. The plasmin system is induced by and degrades amyloid β aggregates. J Neurosci 2000; 20:3937-46.

56. Riemann D, Hansen GH, Niels-Christiansen L et al. Caveolae/lipid rafts in fibroblast-like synoviocytes: ectopeptidase- rich membrane microdomains. Biochem J 2001; 354:47-55.

57. Ledesma MD, Da Silva JS, Crassaerts K et al. Brain plasmin enhances APP α-cleavage and Aβ degradation and is reduced in Alzheimer's disease brains. EMBO Rep 2000; 1:530-5.

58. Wolozin B, Kellman W, Ruosseau P et al. Decreased prevalence of Alzheimer disease associated with 3-hydroxy-3- methyglutaryl coenzyme A reductase inhibitors. Arch Neurol 2000; 57:1439-43.

59. Notkola IL, Sulkava R, Pekkanen J et al. Serum total cholesterol, apolipoprotein E epsilon 4 allele, and Alzheimer's disease. Neuroepidemiology 1998; 17:14-20.

60. Jarvik GP, Wijsman EM, Kukull WA et al. Interactions of apolipoprotein E genotype, total cholesterol level, age, and sex in prediction of Alzheimer's disease: a case-control study. Neurology 1995; 45:1092-6.

CHAPTER 9

Amyloid β-Protein in Low-Density Membrane Domains

Maho Morishima-Kawashima and Yasuo Ihara

Abstract

Membrane microdomains, which have just been emerging thanks to recent technological progress, may have significant roles in normal cell functions such as adhesion, signaling, and trafficking. We found that a significant amount of amyloid β-protein (Aβ) is located to the cholesterol- and sphingolipid-rich low-density membrane (LDM) microdomains, and that these by themselves can produce Aβ. When $A\beta_{42}$ starts to accumulate in the brain, the levels of $A\beta_{42}$ are invariably also increased in LDM domains. The levels of $A\beta_{42}$ in LDM domains correlate well with the extent of extracellular Aβ deposition. In addition, mutant presenilin appears to recruit a larger amount of $A\beta_{42}$ to LDM domains. Thus, the $A\beta_{42}$ associated with LDM domains may play a significant role in the Aβ deposition that is usually observed in aged human brains.

Introduction

Senile plaques and neurofibrillary tangles are two neuropathological hallmarks of Alzheimer's disease (AD). The latter are composed of a mid-molecular weight microtubule-associated phosphoprotein, tau, and they accumulate within neurons. The former are composed of amyloid β-protein (Aβ), a small hydrophobic peptide of unknown function which accumulates in the extracellular space of the brain. Deposition of Aβ is seen more than a decade before neurofibrillary tangle formation occurs in the neocortex in the brains of Down's syndrome patients, who invariably develop AD pathology by their thirties to forties.[1] All three causative and two susceptibility genes for AD thus far known enhance the production or accumulation of $A\beta_{42}$, a more amyloidogenic Aβ species.[2-6] Thus, the accumulation of $A\beta_{42}$ in the brain is a likely initial requisite step for the development of AD.

Aβ is produced from β-amyloid precursor protein (APP) through two sequential steps of proteolysis. β-secretase, which was identified as BACE1 (β-site APP cleaving enzyme) in 1999,[7-10] cleaves at the N-terminus of Aβ, generating the 99-residue-β-carboxy-terminal fragment (CTF). βCTF is then cleaved by γ-secretase, mainly after the Val40 and Ala42 amino acids, releasing $A\beta_{40}$ and $A\beta_{42}$, respectively. It is generally agreed that γ-secretase activity is presenilin (PS)-dependent, but whether γ-secretase is PS itself is still a matter of debate.[11] Neither is it yet known where Aβ is produced within the cell. Currently, most of the Aβ produced within the cell seems to be secreted into the extracellular fluid, which explains the presence of Aβ in the cerebrospinal fluid and plasma in normal subjects. $A\beta_{40}$ is the predominant species (~90%) among the secreted Aβ forms, as is also shown by a cell-free Aβ generation system.[12] In contrast, it is $A\beta_{42}$ that initially and then predominantly is deposited in the brain.[13,14] This may be explained by the much higher aggregation potential of $A\beta_{42}$ compared with the other species.[15] However, why and how $A\beta_{42}$ accumulates and aggregates in the brain parenchyma is still an unresolved issue.

Age-Dependent Aβ Accumulation and Deposition in the Human Brain

Aβ deposition as senile plaques is observed in human brains affected by only a limited number of diseases, including AD. However, such senile plaques are also observed in most brains of aged nondemented subjects. Accurate Aβ quantification by sandwich enzyme-linked immunosorbent assays (ELISA) has shed some light on the underlying process.[5,14] Aβ accumulates in the insoluble fraction of brain homogenates in an age-dependent manner (Fig. 9.1), which supports the view that aging is a strong risk factor for developing AD. The presence of insoluble Aβ itself is not abnormal because it is found even in the brains of nondemented subjects as young as 20-30 years old. In the insoluble fraction from these young brains, the levels of $A\beta_{40}$ are always several- to 10-fold higher than those of $A\beta_{42}$, a proportion corresponding with that of the secreted Aβ.

Our cross-sectional data on Aβ levels in brains from many nondemented subjects aged 20-80 years can be reconstituted into a possible longitudinal profile with some reservations.[5] The initially low $A\beta_{40/42}$ levels (below 5 pmol/g) seem to be stable in individuals aged 20-40, but then start to increase exponentially in some individuals. The incidence of significant Aβ accumulation in brains increases age-dependently. The levels of $A\beta_{40}$ and $A\beta_{42}$ are well correlated and increase in a coordinated manner, but the increasing rate with age for $A\beta_{42}$ is much steeper than that for $A\beta_{40}$. As a result, $A\beta_{42}$ appears to be the first species deposited and is by far the predominant species in most brains of elderly subjects. This underlies the immunohistochemical observations that the appearance of $A\beta_{42}$-positive plaques precedes that of $A\beta_{40}$-positive ones.[13] Accumulation of $A\beta_{42}$ reaches seemingly a plateau (~10,000 pmol/g) at around 70 years of age, while $A\beta_{40}$ tends to continue to increase (Fig. 9.1). In other words, $A\beta_{40}$ continues to increase even after "saturation" of $A\beta_{42}$. This may explain why, in terms of Aβ deposition, the major characteristic of the brains of AD patients is reported to be an increased level of $A\beta_{40}$, but not $A\beta_{42}$,[14,16] and why apolipoprotein E (ApoE) ε4 allele in these individuals is apparently associated with a greater number of $A\beta_{40}$-positive plaques and increased levels of $A\beta_{40}$[17,18] One can surmise that the levels of $A\beta_{40}$ represent the duration of Aβ accumulation: higher levels of $A\beta_{40}$ presumably reflect a longer history of Aβ deposition in any particular individual's brain.

Insoluble Aβ Is Located in LDM Domains

During our study on intracellular Aβ in human neuroblastoma SH-SY5Ycells, we found that a substantial fraction of Aβ within the cell is resistant to extraction with 1% Triton X-100.[19] More than two-thirds of the intracellular $A\beta_{40}$ and $A\beta_{42}$ are left in the Triton-insoluble fraction. This was also the case with NT2N cells, in which insoluble $A\beta_{42}$ was reported to increase during neuronal differentiation induced by retinoic acid.[20]

This Triton X-100 insolubility brings to mind the detergent-insoluble, glycolipid-enriched domains (DIGs).[21,22] These lipids constitute the liquid-ordered phase within the liquid-disordered phase of the membrane. DIGs were previously named caveolae after their characteristic structure formed by invagination from the plasma membrane. They are currently often referred to as lipid rafts, assuming them to be platforms moving within the membrane.[21] Most interestingly, cholesterol and ganglioside GM1, which are abundant in and specific for the microdomains, bind to Aβ molecules and induce them into forming fibrils.[23-25] DIGs are probably involved in transporting various proteins and lipids, in signal transduction, cell adhesion, and mediating cell polarity.[21,22] These DIG functions might be interfered by the process of Aβ accumulation or aggregation. Thus, we examined whether the Triton-insoluble Aβ

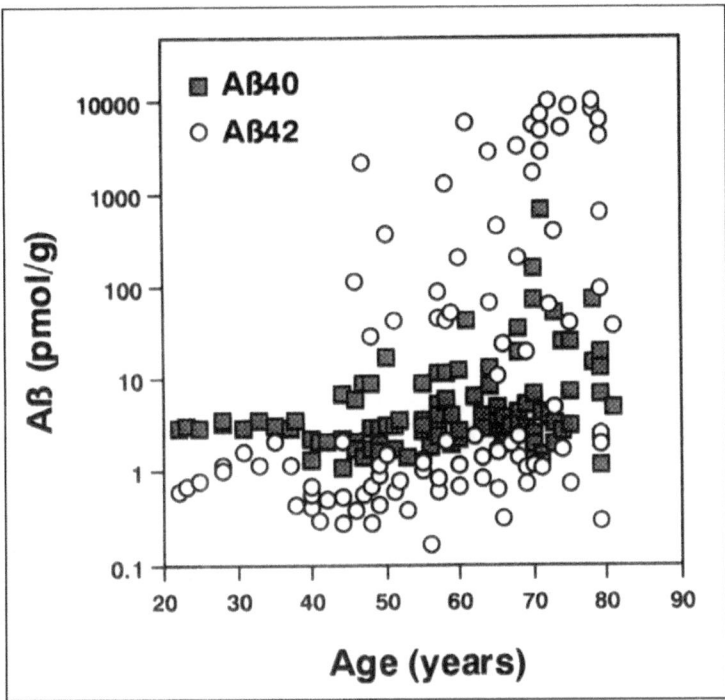

Figure 9.1. Aβ levels in the insoluble fraction of brains from nondemented subjects of various ages. Aβ40 (shadowed squares) and Aβ42 (open circles) were extracted from the Tris-saline insoluble pellets of 105 nondemented individuals' brains with 6 M guanidine hydrochloride and quantified by two-site ELISA. The values obtained are plotted against age at death. The incidence of significantly Aβ-accumulated subjects increases in an age-dependent manner. Note that the *y* axis is a logarithmic scale, and thus Aβ42 levels increase much more steeply than those of Aβ40.

represents an association with a specific membrane domains. Because of their specific lipid composition, DIGs can be readily isolated as a low-density fraction by floating up through sucrose density gradients in the presence of Triton X-100.[26,27] However, recent studies indicate the presence of multiple subtypes of microdomains in the low-density membrane fraction.[28,29] Thus, the membrane component fractionated into the buoyant fraction is referred to here to as low-density membrane: LDM.

Following sucrose density gradient centrifugation of the homogenates prepared from SH-SY5Y cells, approximately half of the Triton X-100 insoluble Aβ40 and Aβ42 was recovered in the LDM fraction.[19] Similar observations were reported by Lee et al[30] using rat brains and APP-overexpressing Chinese hamster ovary (CHO) cells. In addition, they showed by immunoelectron microscopy the association of Aβ with isolated caveolae-like membrane vesicles in the LDM fraction.[30] These indicate a specific association of at least a fraction of the intra-cellular Aβ with LDM domains. Because of much smaller amounts of proteins in LDM do-mains, Aβ appears to be highly concentrated in the membrane domains in terms of protein. Thus, it is possible that LDM domains are a particular compartment that initially accumulates Aβ, because aggregation of Aβ is concentration-dependent and its physiological concentra-tions are usually under the critical concentration, and because aggregation-promoting lipid molecules are enriched in the domains (see below).

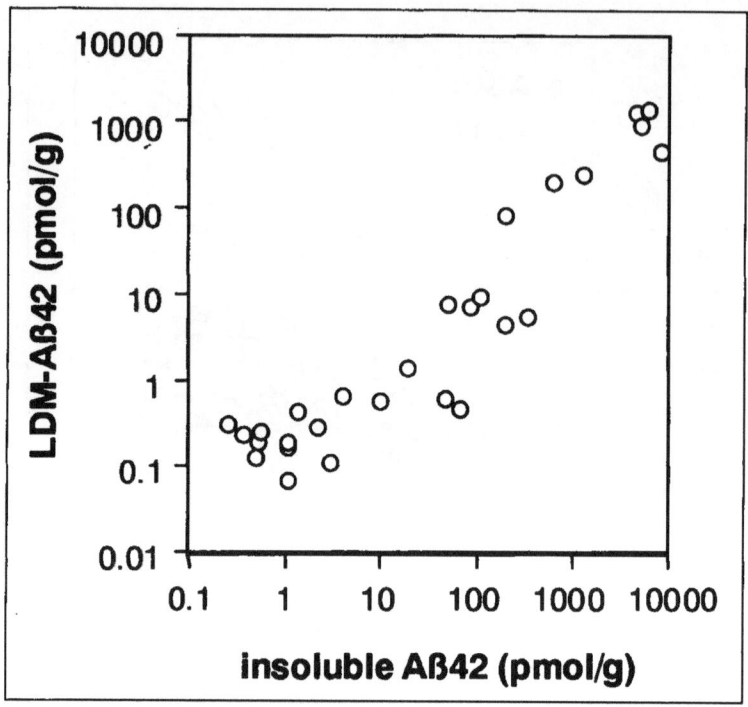

Figure 9.2. Relation between Aβ42 levels in LDM domains (LDM-Aβ42; *y* axis) and insoluble Aβ42 levels (*x* axis). Small cortical blocks from autopsied brains were homogenized in the presence of 1% Triton X-100 and fractionated by sucrose density gradient centrifugation. The low-density fraction was collected and spun down. The resulting pellet was extracted with 6 M guanidine hydrochloride, and the extract was subjected to ELISA for Aβ. The levels of Aβ42 in LDM domains are proportional to the total insoluble Aβ42 levels, which in turn correlate with amyloid burden.

Aβ Accumulation in LDM Domains Accurately Reflects the Extent of Aβ Deposition in the Brain

To investigate the origin of insoluble Aβ in the human brain and to examine whether Aβ in LDM domains of the brain is involved in the pathological cascade leading to AD, Aβ levels in the membrane domains from many nondemented subjects of various ages were quantified.[31] Homogenates from many human brains were fractionated by sucrose density gradient centrifugation, and the contents of Aβ in the LDM fractions were quantified using a two-site ELISA. A substantial fraction of the insoluble Aβ was associated with LDM fractions, even in the brains from nondemented young subjects that did not exhibit significant Aβ accumulation. The levels of $Aβ_{40}$ were several-fold higher than those of $Aβ_{42}$ in the LDM fractions from young brains. During aging, LDM fractions appeared to preferentially accumulate $Aβ_{42}$. Moreover, the levels of $Aβ_{42}$ in those fractions were proportional to total insoluble $Aβ_{42}$ levels (Fig. 9.2), which is known to correlate well with amyloid burden.[14]

To exclude the possibility that this association comes from post-mortem alterations, which often complicate human samples, we examined the PDAPP strain of mice, which overexpresses APPV717F.[31,32] These mice develop senile plaques by the age of four months and progressively accumulate Aβ in a manner similar to human brains.[33] Similar $Aβ_{42}$ accumulation in LDM fractions was observed in the mice as Aβ accumulation progressed with aging (Fig. 9.3). LDM

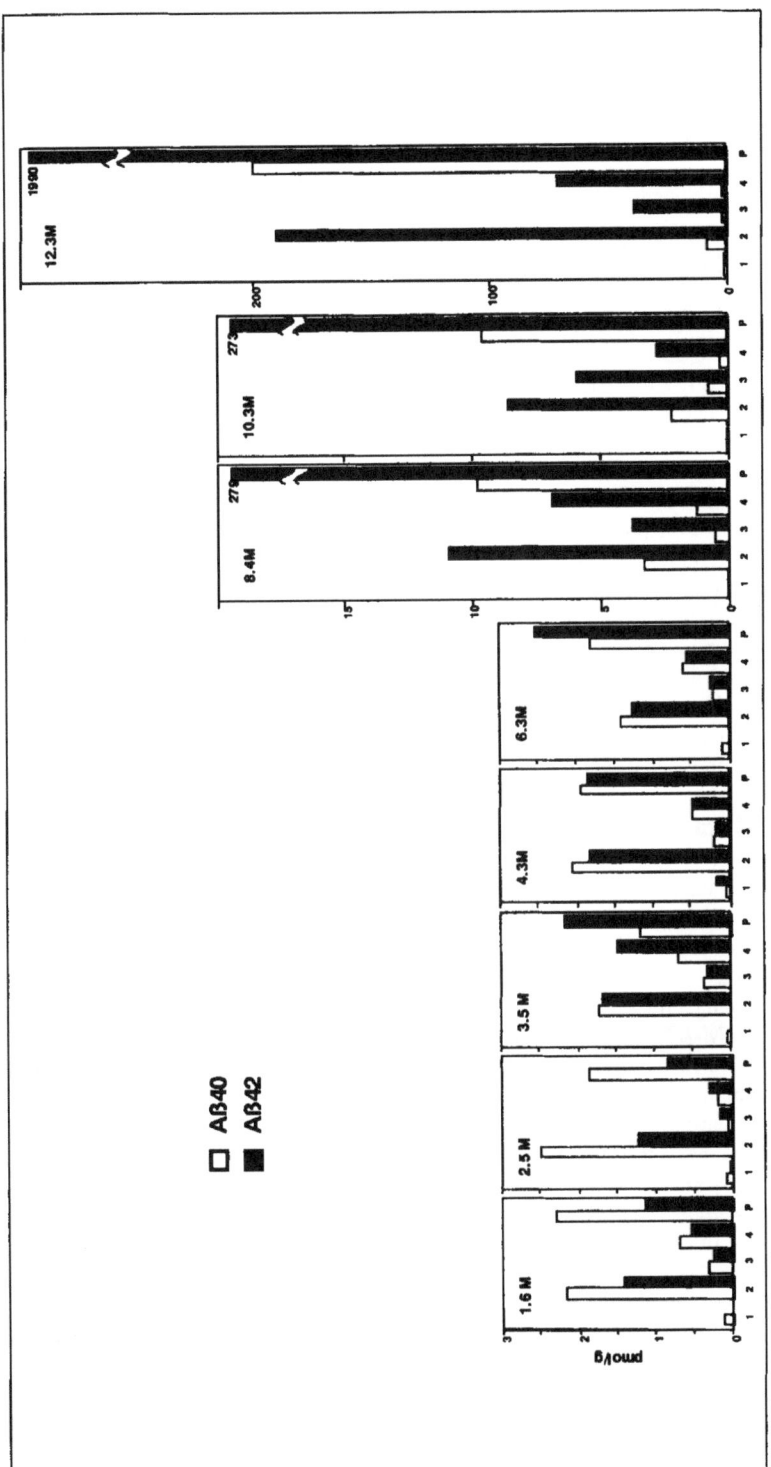

Figure 9.3. Sucrose density gradient fractionation of brain homogenates from PDAPP mice. Brains from PDAPP mice aged 1.6 to 12.3 months were homogenized and fractionated by sucrose density gradient centrifugation in the presence of 1% Triton X-100. Each fraction collected from the top was centrifuged and the resulting pellet was extracted with guanidine hydrochloride. The extract was subjected to ELISA for Aβ40 (open bars) and Aβ42 levels (closed bars). The Aβ42 levels in the pellet (P) reflect insoluble Aβ deposition. In PDAPP mice, the accumulation of Aβ42 in LDM domains (fraction 2) increases as Aβ deposition progresses with aging.

fractions prepared from mouse or human brain were always contaminated with myelin-derived creamy material, raising the possibility that a fraction of Aβ is bound to myelin during homogenization and floats to the LDM fraction. Exogenous Aβ added to the homogenates, however, was not fractionated in the LDM fraction. Additionally, another fractionation protocol separated Triton X-100-insoluble Aβ from a myelin marker protein. Thus, the Aβ that was fractionated into the LDM fraction appears to represent largely the Aβ associated with LDM domains (LDM-Aβ).

These observations suggest that LDM-Aβ$_{42}$ is coupled with extracellular Aβ deposition and that one pathway to Aβ$_{42}$ accumulation in the brain is mediated through an abnormal accumulation of Aβ$_{42}$ in specific membrane domains. Consistent with this assumption, careful immunoelectron microscopic studies showed that Aβ accumulates on juxta-plasma membranes at the very early stage of Aβ deposition in AD brains.[34]

We attempted to extract the Aβ in LDM domains with various kinds of detergents and solvents. Unexpectedly, LDM-Aβ was not solubilized by various kinds of detergents, including 1% Triton X-100 at 37 °C, 1% octyl β-D-glucoside, 1% CHAPS, 1% Sarkosyl, and 0.5% SDS. Of the solvents examined thus far, only 0.1 M sodium carbonate extracted some LDM-Aβ. These results suggest that LDM-Aβ is tightly bound to the membrane, presumably on the membrane surface. Notably, octyl β-D-glucoside is known to disrupt the caveolae-like membranes and dissociate resident proteins from the membrane domains.[35] Consistent with the above observation on the human brain, the Aβ in the Triton-insoluble fraction of SY5Ycells cannot be extracted with octyl β-D-glucoside. Thus, our results may indicate that LDM-Aβ is associated with the microdomains distinct from well-characterized caveolae-like membranes. Another possibility is that a specific resident protein(s) may interact with and anchor Aβ in LDM domains. We sought to find such a particular protein, the level of which is increased according to Aβ$_{42}$ accumulation in LDM domains, but have so far failed to find it. Interaction of Aβ with membrane lipids would be a third possibility (see below).

Effects of Mutant Presenilin on the Aβ in LDM Domains

Familial AD (FAD)-associated mutations of PS1/2 are known to increase the production of Aβ$_{42}$,[3,4] which likely leads to earlier accumulation and deposition of Aβ$_{42}$ in the brain.[36] Thus, we next assessed whether mutant (mt) PS has an effect on the Aβ$_{42}$ in LDM domains. Analysis of transgenic mice clearly showed that mtPS2, but not wild-type (wt) PS2, greatly enhanced the levels of Aβ$_{42}$ in LDM domains (Fig. 9.4).[37] An increase of LDM-Aβ$_{42}$ was more remarkable in younger mice, compared with age-matched wtPS2- or nontransgenic mice. Such an increase in LDM-Aβ$_{42}$ was also observed in CHO cells stably overexpressing N141I mtPS2. Because the mtPS2-transgenic mice appear to recapitulate the initial phase of Aβ accumulation in the human brain, these observations suggest that LDM domains are one of the cellular locations accumulating Aβ$_{42}$.

Another interesting feature was that the levels of sphingomyelin and glycerophospholipids in LDM domains were increased in some lines of mtPS2-transgenic mice, compared with wtPS2-transgenic ones.[37] Although further studies are required, the increased levels of LDM-Aβ$_{42}$ may be mediated through the altered lipid composition induced by mtPS2.

Aβ Production by LDM Domains

LDM domains contain essential components of γ-secretase complex including βCTF of APP, and N-terminal fragments (NTF) and CTF of PS, and were postulated to be one of the locations for Aβ production.[30] Recently, it was reported that the buoyant cholesterol-rich membrane microdomains, which may correspond to LDM domains defined here, have high γ-secretase activity.[38] Moreover, the activity depends on the cholesterol content of the

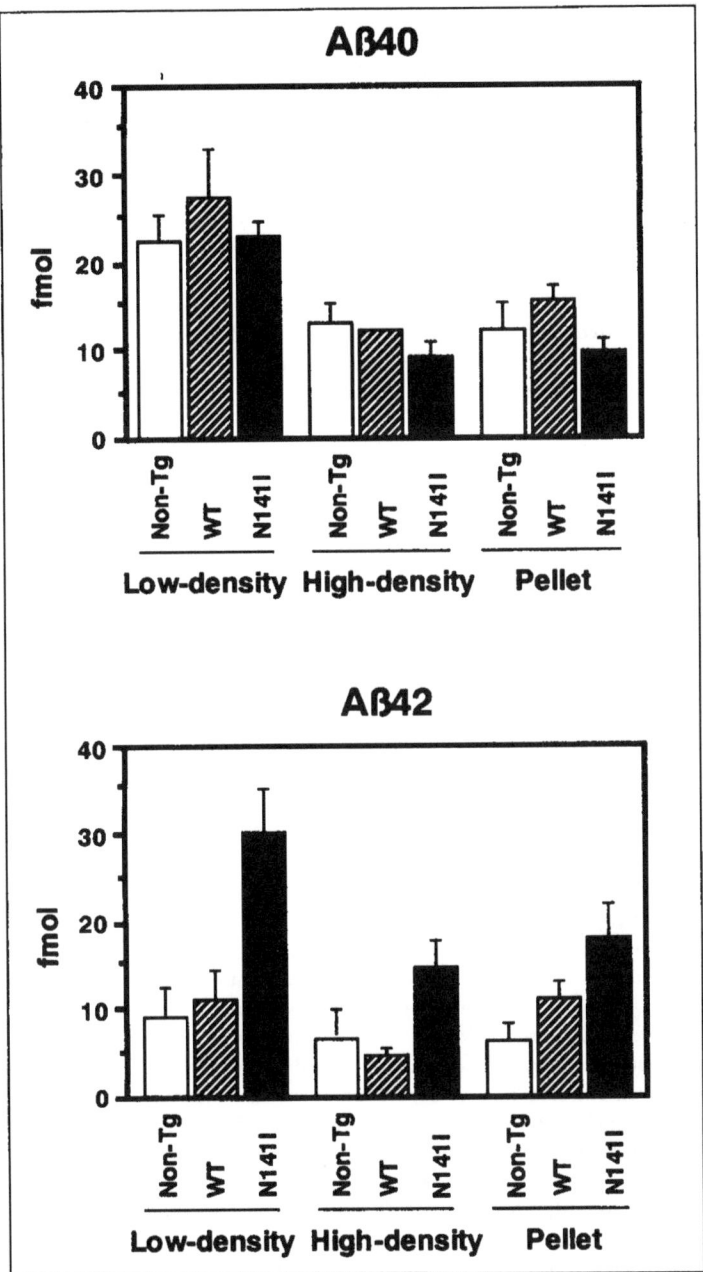

Figure 9.4. Fractionation of the insoluble Aβ in brains from nontransgenic (Non-Tg), wild-type (WT) and N141I mutant (N141I) PS2-transgenic mice. Each brain from two-month-old Non-Tg or transgenic mice was homogenized in the presence of Triton X-100 and fractionated by sucrose density gradient centrifugation. After brief centrifugation of the collected fractions, the pellets were extracted with guanidine hydrochloride, and the extracts were subjected to ELISA for Aβ40 (upper panel) and Aβ42 levels (lower panel).

membrane.[38] Thus, we sought to assess γ-secretase activities of LDM domains using our cell-free Aβ production system.

We used either CHAPS or CHAPSO instead of Triton X-100 to prepare LDM fractions from CHO cells overexpressing APP. This is because γ-secretase activity is susceptible to Triton X-100 and can be retained at least in part with the former two.[12] Although sodium carbonate, previously used for isolation of DIGs,[39] can keep γ-secretase active, the LDM fractions obtained were contaminated by proteins derived from other organelles, such as the endoplasmic reticulum and Golgi apparatus. Following sucrose density gradient centrifugation, we found that CHAPSO enabled flotillin and caveolin—marker proteins for DIGs—to fractionate preferentially and almost exclusively into the buoyant fraction. Cholesterol was abundant in the LDM fraction, and APP, α and βCTF, and PS fragments were present. Incubation of the LDM fraction thus prepared produced significant amounts of Aβ. This indicates that LDM domains (or more accurately, a certain subgroup of microdomains fractionated in the LDM fraction) are most likely one of the compartments producing Aβ.

The content of cholesterol in the membrane affects the distribution and activities of many raft-specific proteins.[40] Lowering cholesterol content suppresses the formation of cholesterol-rich microdomains, and normally raft-resident proteins are no longer localized to rafts. Cholesterol itself is involved in the thickness of the membrane and helps keep it rigid. Thus, it would be reasonable to speculate that lowering cholesterol content might produce significant effects on the distribution of the γ-secretase complex composed of several transmembrane proteins, its intramembranous topology and, most importantly, its activity. Our preliminary data, however, showed that depletion of cholesterol from the membrane by methyl-β-cyclodextrin caused no significant decrease in the production of $Aβ_{40}$ and $Aβ_{42}$ despite a substantial reduction in the content of membrane cholesterol. Moreover, in our hands depletion of cholesterol from the total cellular membranes did not affect Aβ production. Thus, the levels of membrane cholesterol appear to have no effect on the γ-secretase activity. The methyl-β-cyclodextrin treatment decreases cholesterol levels and disrupts the liquid ordered phase of the membrane.[40] Nonetheless, the γ-secretase activity and βCTF, its substrate did not dissociate from, but kept their association with the membrane. This is surprising and needs a plausible explanation. One possibility is that the γ-secretase activity is associated with particular microdomains distinct from those depleted of cholesterol. In either way, our results clearly conflict with those obtained by the other group.[38]

Cholesterol is currently receiving much attention, because recent epidemiological studies suggested that statins, cholesterol-lowering agents that inhibit HMG CoA reductase, decrease the prevalence of AD.[41,42] Concomitantly, many investigators have reported that cholesterol levels are involved in Aβ production.[43-47] Overall, an increase in cholesterol levels appears to be associated with an increased accumulation of Aβ and vice-versa.[43,45-47] However, most of those studies used cultured cells or experimental animals and did not necessarily provide a direct link between cholesterol levels and Aβ production. Manipulation of the cellular cholesterol levels will induce alterations in the levels of other lipids as well as its distribution among the organelles.[48] These may further alter intracellular transport and impair membrane trafficking, which may affect the distribution of various membrane proteins including γ-secretase complex and its substrates.[49] It is also possible that those agents have multiple targets other than cholesterol levels, one of which may influence Aβ production. For example, nonsteroidal anti-inflammatory drugs (NSAIDs) have been known to inhibit cyclooxygenase, but a subset of them suppresses the production of $Aβ_{42}$ as well.[50] Thus, although the above studies imply an intimate relationship between Aβ production and membrane lipids, especially cholesterol, the underlying mechanisms remain to be elucidated.

Possible Involvement of LDM-Aβ in Aβ Accumulation in the Brain

From the above data, it is plausible to assume that accumulation of LDM-Aβ$_{42}$ leads to its aggregation and formation of fibrils in LDM domains. In fact, amyloid fibrils were observed electron microscopically in the LDM fraction prepared from PDAPP mouse brains.[51]

Aβ is known to bind ganglioside GM1.[24] This interaction is markedly enhanced by the presence of cholesterol, which induces gangliosides to cluster together and helps Aβ molecules assume a β-sheet structure, leading to fibril formation.[52-54] The Aβ$_{42}$ produced in LDM domains may be efficiently turned over but tends to accumulate, possibly due to an age-related slowing of its turnover rate. Finally, it initiates aggregations surrounded by cholesterol and GM1 molecules. Consistent with this assumption, most of the Aβ produced in the cell-free system is tightly bound to the membrane and cannot be released into the reaction mixture. In the brains of individuals affected by FAD-linked mtPS, production of Aβ$_{42}$ by LDM domains may be increased to only a small extent but from the very initial stage, causing increased levels of LDM-Aβ$_{42}$, as observed in the mtPS2-transgenic mice.[37] Thus, Aβ$_{42}$-concentrated microdomains are a strong candidate as the site of the initial Aβ$_{42}$ aggregation.

However, an increase in Aβ$_{42}$ production has never been known for individuals with sporadic AD. In this case, slowing of the clearance rate of Aβ$_{42}$ and/or alterations in the membrane lipid composition, which presumably occur during aging,[55,56] may be involved in the Aβ$_{42}$ aggregation in LDM domains. In this context, it would be particularly important to learn the effects of ApoE alleles on LDM-Aβ, because ε4 is a strong risk factor for developing AD[57,58] and its presence is tightly associated with an earlier accumulation of insoluble Aβ$_{42}$ in the human brain.[5] Thus, we quantified the Aβ levels in the LDM fraction from human ApoE3 and ApoE4 knockin mouse brains.[59] So far, no significant difference has been observed in the LDM-Aβ$_{40}$/Aβ$_{42}$ levels up to 12 months between humanized ApoE3 and ApoE4 mice. The accumulation of LDM-Aβ$_{42}$ may thus not be mediated through ApoE4 or its interacting proteins.

Altogether, the current studies suggest a possible involvement of LDM-Aβ in the initial and late phase of Aβ deposition, but further studies are required to determine its pathological significance.

References

1. Mann DM. Cerebral amyloidosis, ageing and Alzheimer's disease; a contribution from studies on Down's syndrome. Neurobiol Aging 1989; 10:397-399.
2. Suzuki N, Cheung TT, Cai XD et al. An increased percentage of long amyloid β protein secreted by familial amyloid β protein precursor (βAPP717) mutants. Science 1994; 264:1336-1340.
3. Scheuner D, Eckman C, Jensen M et al. Secreted amyloid β-protein similar to that in the senile plaques of Alzheimer's disease is increased in vivo by the presenilin 1 and 2 and APP mutations linked to familial Alzheimer's disease. Nat Med 1996; 2:864-870.
4. Borchelt DR, Thinakaran G, Eckman CB et al. Familial Alzheimer's disease-linked presenilin 1 variants elevate Aβ$_{1-42/1-40}$ ratio in vitro and in vivo. Neuron 1996; 17:1005-1013.
5. Morishima-Kawashima M, Oshima N, Ogata H et al. Effect of apolipoprotein E allele ε4 on the initial phase of amyloid β-protein accumulation in the human brain. Am J Pathol 2000; 157:2093-2099.
6. Ertekin-Taner N, Graff-Radford N, Younkin LH et al. Linkage of plasma Aβ42 to a quantitative locus on chromosome 10 in late-onset Alzheimer's disease pedigrees. Science 2000; 290:2303-2304.
7. Vassar R, Bennett BD, Babu-Khan S et al. β-secretase cleavage of Alzheimer's amyloid precursor protein by the transmembrane aspartic protease BACE. Science 1999; 286:735-741.
8. Yan R, Bienkowski MJ, Shuck ME et al. Membrane-anchored aspartyl protease with Alzheimer's disease β-secretase activity. Nature 1999; 402:533-537.
9. Sinha S, Anderson JP, Barbour R et al. Purification and cloning of amyloid precursor protein β-secretase from human brain. Nature 1999; 402:537-540.

10. Hussain I, Powell D, Howlett DR et al. Identification of a novel aspartic protease (Asp 2) as β-secretase. Mol Cell Neurosci 1999; 14:419-427.

11. Selkoe DJ. Alzheimer's disease: genes, proteins, and therapy. Physiol Rev 2001; 81:741-766.

12. Li YM, Lai MT, Xu M et al. Presenilin 1 is linked with γ-secretase activity in the detergent solubilized state. Proc Natl Acad Sci USA 2000; 97:6138-6143.

13. Iwatsubo T, Odaka A, Suzuki N et al. Visualization of Aβ$_{42(43)}$ and Aβ$_{40}$ in senile plaques with end-specific Aβ monoclonals: evidence that an initially deposited species is Aβ$_{42 (43)}$. Neuron 1994; 13:45-53.

14. Funato H, Yoshimura M, Kusui K et al. Quantitation of amyloid β-protein (Aβ) in the cortex during aging and in Alzheimer's disease. Am J Pathol 1998; 152:1633-1640.

15. Jarrett JT, Berger EP, Lansbury PT Jr. The carboxy terminus of the β amyloid protein is critical for the seeding of amyloid formation: implications for the pathogenesis of Alzheimer's disease. Biochemistry 1993; 32:4693-4697.

16. Shinkai Y, Yoshimura M, Morishima-Kawashima M et al. Amyloid β-protein deposition in the leptomeninges and cerebral cortex. Ann Neurol 1997; 42:899-908.

17. Gearing M, Mori H, Mirra SS. Aβ-peptide length and apolipoprotein E genotype in Alzheimer's disease. Ann Neurol 1997; 39:395-399.

18. Ishii K, Tamaoka A, Mizusawa H, et al. Aβ$_{1-40}$ but not Aβ$_{1-42}$ levels in cortex correlate with apolipoprotein E ε4 allele dosage in sporadic Alzheimer's disease. Brain Res 1997; 748:250-252.

19. Morishima-Kawashima M, Ihara Y. The presence of amyloid β-protein in the detergent-insoluble membrane compartment of human neuroblastoma cells. Biochemistry 1998; 37:15247-15253.

20. Skovronsky DM, Doms RW, Lee VM. Detection of a novel intraneuronal pool of insoluble amyloid β protein that accumulates with time in culture. J Cell Biol 1998; 141:1031-1039.

21. Simons K, Ikonen E. Functional rafts in cell membranes. Nature 1997; 387:569-572.

22. Brown DA, London E. Functions of lipid rafts in biological membranes. Annu Rev Cell Dev Biol 1998; 14:111-136.

23. Avdulov NA, Chochina SV, Igbavboa U et al. Lipid binding to amyloid β-peptide aggregates: preferential binding of cholesterol as compared with phosphatidylcholine and fatty acids. J Neurochem 1997; 69:1746-1752.

24. Choo-Smith LP, Surewicz WK. The interaction between Alzheimer amyloid β(1-40) peptide and ganglioside GM1-containing membranes. FEBS Lett 1997; 402: 95-98.

25. Choo-Smith LP, Garzon-Rodriguez W, Glabe CG et al. Acceleration of amyloid fibril formation by specific binding of Aβ-(1-40) peptide to ganglioside-containing membrane vesicles. J Biol Chem 1997; 272:22987-22990.

26. Brown DA, Rose JK. Sorting of GPI-anchored proteins to glycolipid-enriched membrane subdomains during transport to the apical cell surface. Cell 1992; 68:533-544.

27. Lisanti MP, Scherer PE, Vidugiriene J et al. Characterization of caveolin-rich membrane domains isolated from an endothelial-rich source: implications for human disease. J Cell Biol 1994; 126:111-126.

28. Gomez-Mouton C, Abad JL, Mira E et al. Segregation of leading-edge and uropod components into specific lipid rafts during T cell polarization. Proc Natl Acad Sci USA 2001; 98:9642-9647.

29. Hansen GH, Immerdal L, Thorsen E et al. Lipid rafts exist as stable cholesterol-independent microdomains in the brush border membrane of enterocytes. J Biol Chem 2001; 276:32338-32344.

30. Lee SJ, Liyanage U, Bickel PE et al. A detergent-insoluble membrane compartment contains Aβ in vivo. Nat Med 1998; 4:730-734.

31. Oshima N, Morishima-Kawashima M, Yamaguchi H et al. Accumulation of amyloid β-protein in the low-density membrane domain accurately reflects the extent of β-amyloid deposition in the brain. Am J Pathol 2001; 158:2209-2218.

32. Games D, Adams D, Alessandrini R et al. Alzheimer-type neuropathology in transgenic mice overexpressing V717F β-amyloid precursor protein. Nature 1995; 373:523-527.

33. Johnson-Wood K, Lee M, Motter R et al. Amyloid precursor protein processing and Aβ$_{42}$ deposition in a transgenic mouse model of Alzheimer disease. Proc Natl Acad Sci USA 1997; 94:1550-1555.

34. Yamaguchi H, Maat-Schieman ML, van Duinen SG et al. Amyloid β protein (Aβ) starts to deposit as plasma membrane-bound form in diffuse plaques of brains from hereditary cerebral hemorrhage with amyloidosis-Dutch type, Alzheimer disease and nondemented aged subjects. J Neuropathol Exp Neurol 2000; 59:723-732.

35. Schroeder R, London E, Brown D. Interactions between saturated acyl chains confer detergent resistance on lipids and glycosylphosphatidylinositol (GPI)-anchored proteins: GPI-anchored proteins in liposomes and cells show similar behavior. Proc Natl Acad Sci USA 1994; 91:12130-12134.

36. Lemere CA, Lopera F, Kosik KS et al. The E280A presenilin 1 Alzheimer mutation produces increased Aß42 deposition and severe cerebellar pathology. Nat Med 1996; 2:1146-1150.

37. Sawamura N, Morishima-Kawashima M, Waki H et al. Mutant presenilin 2 transgenic mice. A large increase in the levels of $A\beta_{42}$ is presumably associated with the low density membrane domain that contains decreased levels of glycerophospholipids and sphingomyelin. J Biol Chem 2000; 275:27901-27908.

38. Wahrle S, Das P, Nyborg AC et al. Cholesterol-dependent γ-secretase activity in buoyant cholesterol-rich membrane microdomains. Neurobiol Dis 2002; 9:11-23.

39. Song KS, Li Shengwen, Okamoto T et al. Co-purification and direct interaction of Ras with caveolin, an integral membrane protein of caveolae microdomains. Detergent-free purification of caveolae microdomains. J Biol Chem 1996; 271:9690-9697.

40. Simons K, Toomre D. Lipid rafts and signal transduction. Nat Rev Mol Cell Biol 2000; 1:31-39.

41. Jick H, Zornberg GL, Jick SS et al. Statins and the risk of dementia. Lancet 2000; 356:1627-1631.

42. Wolozin B, Kellman W, Ruosseau P et al. Decreased prevalence of Alzheimer disease associated with 3-hydroxy-3-methylglutaryl coenzyme A reductase inhibitors. Arch Neurol 2000; 57:1439-1443.

43. Simons M, Keller P, De Strooper B et al. Cholesterol depletion inhibits the generation of β-amyloid in hippocampal neurons. Proc Natl Acad Sci USA 1998; 95:6460-6464.

44. Howland DS, Trusko SP, Savage MJ et al. Modulation of secreted β-amyloid precursor protein and amyloid β-peptide in brain by cholesterol. J Biol Chem 1998; 273:16576-16582.

45. Frears ER, Stephens DJ, Walters CE et al. The role of cholesterol in the biosynthesis of β-amyloid. Neuroreport 1999; 10:1699-1705.

46. Refolo LM, Malester B, LaFrancois J et al. Hypercholesterolemia accelerates the Alzheimer's amyloid pathology in a transgenic mouse model. Neurobiol Dis 2000; 7:321-331.

47. Fassbender K, Simons M, Bergmann C et al. Simvastatin strongly reduces levels of Alzheimer's disease β-amyloid peptides $A\beta_{42}$ and $A\beta_{40}$ in vitro and in vivo. Proc Natl Acad Sci USA 2001; 98:5856-5861.

48. Simons K, Ikonen E. How cells handle cholesterol. Science 2000; 290:1721-1726.

49. Runz H, Rietdorf J, Tomic I et al. Inhibition of intracellular cholesterol transport alters presenilin localization and amyloid precursor protein processing in neuronal cells. J Neurosci 2002; 22:1679-1689.

50. Weggen S, Eriksen JL, Das P et al. A subset of NSAIDs lower amyloidogenic $A\beta_{42}$ independently of cyclooxygenase activity. Nature 2001; 414:212-216.

51. Kawarabayashi T, Shoji M, Wahrle S et al. Amyloid β protein accumulates in lipid rafts. Soc Neurosci Abstr 2001; 27:583.3.

52. Matsuzaki K, Horikiri C. Interactions of amyloid β-peptide (1-40) with ganglioside-containing membranes. Biochemistry 1999; 38:4137-4142.

53. Kakio A, Nishimoto SI, Yanagisawa K et al. Cholesterol-dependent formation of GM1 ganglioside-bound amyloid β-protein, an endogenous seed for Alzheimer amyloid. J Biol Chem 2001; 276:24985-24990.

54. Kakio A, Nishimoto S, Yanagisawa K et al. Interactions of amyloid β-protein with various gangliosides in raft-like membranes: importance of GM1 ganglioside-bound form as an endogenous seed for Alzheimer amyloid. Biochemistry 2002; 41:7385-7390.

55. Svennerholm L, Bostrom K, Jungbjer B et al. Membrane lipids of adult human brain: lipid composition of frontal and temporal lobe in subjects of age 20 to 100 years. J Neurochem 1994; 63:1802-1811.

56. Igbavboa U, Avdulov NA, Schroeder F et al. Increasing age alters transbilayer fluidity and cholesterol asymmetry in synaptic plasma membranes of mice. J Neurochem 1996; 66:1717-1725.

57. Strittmatter WJ, Saunders AM, Schmechel D et al. Apolipoprotein E: high-avidity binding to β-amyloid and increased frequency of type 4 allele in late-onset familial Alzheimer disease. Proc Natl Acad Sci USA 1993; 90:1977-1981.

58. Corder EH, Saunders AM, Strittmatter WJ et al. Gene dose of apolipoprotein E type 4 allele and the risk of Alzheimer's disease in late onset families. Science 1993; 261:921-923.

59. Hamanaka H, Katoh-Fukui Y, Suzuki K et al. Altered cholesterol metabolism in human apolipoprotein E4 knock-in mice. Hum Mol Genet 2000; 9:353-361.

CHAPTER 10

Transport-Clearance Hypothesis for Alzheimer's Disease and Potential Therapeutic Implications

Berislav V. Zlokovic and Blas Frangione

Abstract

Alzheimer's disease (AD) is the major cause of dementia and the most common form of human amyloidosis. AD affects an astoundingly large number of people and is common with advancing age. The brains of Alzheimer's patients are typically riddled with insoluble "plaques" which consist of amyloid. The major constituent of brain and cerebrovascular amyloid is amyloid-β peptide (Aβ). A number of genetic, cell biology, biochemical and animal studies support the concept that Aβ is central to AD. Here, we discuss regulation of brain Aβ and propose that amyloidosis in AD is a "storage" disease caused by inefficient transport of this peptide out of the central nervous system. Thus, we suggest that sporadic AD is at least, in part, a clearance disorder due to defects in transport of Aβ that is produced at normal levels throughout the lifetime. The importance of transport-based clearance strategies in conjunction with other Aβ-lowering therapies (e.g., immunization/vaccination, Aβ sequestering agents) in preventing the development of cerebral β-amyloidosis and/or clearing toxic clumps from AD brains is discussed.

Introduction

Alzheimer's disease (AD) is the most common form of human amyloidosis and the major cause of dementia. Alzheimer's is common with advancing age. The brains of Alzheimer's patients are typically riddled with insoluble "plaques" which consists of amyloid—small protein fibers that form a hard mass. Neuropathologically, AD is characterized by:

    i. parenchymal amyloid deposits called neuritic plaques;
    ii. intraneuronal deposits of neurofibrillary tangles;
    iii. cerebral amyloid angiopathy and
    iv. synaptic loss.

The major constituent of the neuritic plaques and congophilic angiopathy is amyloid-β peptide (Aβ).[1] Aβ is 4.3 kD peptide that is produced by proteolytic cleavage from a large type-1 transmembrane protein, the amyloid-β precursor protein (APP).[1]

Genetic Risk Factors for AD

Mutations in three genes encoding amyloid precursor protein (APP) on chromosome 21,[2,3] presenilin 1 on chromosome 14,[4,5] and presenilin 2 on chromosome 1[6,7,8] have been linked to the rare, early onset autosomal form of AD (onset < 60 years). These mutations all affect APP metabolism such that more Aβ is produced. In contrast, sporadic AD that represents most AD cases (> 98%) has age onset typically above 65 years and exhibits no clear pattern of inheritance (late onset AD). The ε4 allele of the apolipoprotein E (ApoE) gene on chromosome 19 appears to be a risk factor for late onset AD.[9,10,11] Recently a susceptibility locus for late onset AD has been identified on chromosome 10[12,13] that acts to increase Aβ levels in plasma in the first-degree relatives of patients with typical late onset AD.[14]

Amyloid β Peptide

Whether Aβ causes or contributes to Alzheimer's dementia is still controversial. A number of genetic, cell biology, biochemical and animal studies support the concept that Aβ plays a central role in the development of AD pathology.[1,15] Elevated cerebral levels of Aβ may occur during normal aging, but the accumulation is significantly accelerated in AD. It is still unresolved how this peptide accumulates in the central nervous system (CNS) and then initiates cytopathology and how much inflammation can contribute to neuronal death. Until recently, significant efforts have been focused on the mechanisms responsible for Aβ production, including the roles of the proteolytic enzymes β- and γ-secretases which generate Aβ from its APP precursor protein.[16,17] Aβ is produced by almost all tissues and cells in the body and circulates in biological fluids such as plasma, cerebrospinal fluid (CSF) and brain interstitial fluid (ISF).[1] It has been suggested that generation of Aβ in the CNS may take place in the neuronal axonal membrane compartment by proteolytic processing of APP after APP-mediated axonal transport of β-secretase and presenilin-1.[18]

Although increases in Aβ production can explain a small percentage of early onset cases of familial AD bearing inherited mutations in APP or presenilins 1 or 2 genes,[19] similar increases in production have not been found in sporadic AD in spite of elevated levels of Aβ. It is believed that minimizing physiological production of Aβ in sporadic AD by blocking β- and γ-secretases may still reduce accumulation of Aβ in the brain. The proof whether Aβ causes Alzheimer's dementia or not should ultimately come from clinical trials with Aβ lowering agents. These trials have recently begun including the vaccination of humans with Aβ and treatments with γ-secretase inhibitors. However, the results of on-going trials at present are inconclusive. The vaccination therapy with Aβ has been halted due to serious side effects, and there is still uncertainty over the feasibility of long-term therapy with secretase inhibitors[16,17] because of their potential effects on other cellular pathways. However, less toxic vaccines and passive immunization are currently under development as well as more selective and specific secretase inhibitors.

A Clearance Mechanism

The realization that increased brain production of Aβ is not involved in sporadic AD or late onset AD led us to propose that amyloidosis in AD can be a "storage" disease caused by inefficient elimination of the peptide from the CNS due to defects in its transport out of the CNS possibly caused by conformational changes in the molecule due to increased concentrations, acidic pH (e.g., inflammation), post-translational modifications (e.g., oxidation, glycation), binding to chaperone molecules and/or down-regulation of the blood-brain barrier (BBB) transporters. The mechanisms of Aβ clearance have received relatively little attention until recently. The two plausible hypotheses for Aβ clearance from the brain are metabolism[20,21] and transport out of the CNS,[22] as we proposed earlier. Here, we will focus on mechanisms for Aβ accumulation caused by inefficient transport of the peptide out of the CNS shortly after its physiological production.

Aβ Transport out of the CNS

The peptide structure of Aβ implies that it cannot be eliminated rapidly from the CNS into circulation unless there is a carrier-mediated and/or receptor-mediated transport system(s) for the peptide in brain microvascular endothelium, a site of the BBB in vivo. Although peptides can be cleared to some extent by passive diffusion or transport via a nonspecific bulk flow of brain ISF and CSF, this route seems to be responsible for about 10% of Aβ clearance from brain tissue under physiological conditions.[23] The BBB has unique properties, as it does not normally allow free exchanges of polar solutes such as Aβ between brain and blood owing to the presence of tight junctions between brain endothelial cells that form a continuous cellular monolayer.

As shown in Figure 10.1, recent studies have revealed a new role for the low density lipoprotein (LDL), receptor related protein-1 (LRP-1), a member of the LDL receptor family which is central to transport and metabolism of cholesterol and ApoE-containing lipoproteins, as a clearance receptor for Aβ at the BBB mediating transport of the peptide from brain into the blood.[23] Based on experimentally determined transport kinetic efflux constants, the LRP-1-mediated transcytosis of Aβ across the BBB is a high affinity and rate limiting step for Aβ clearance from the brain. The LRP-1-mediated Aβ transport is initiated at the abluminal (brain) site of the endothelium and is therefore directly responsible for eliminating Aβ from brain ISF. The rate of transport is modulated by the LRP-1 ligands, ApoE and α2-macroglobulin, that have been identified either as a definite risk factor (i.e., ApoE4)[9-11] and/or a possible risk factor for AD,[24] respectively. Reduced expression of brain endothelial LRP-1 was observed during normal aging in rodents, nonhuman primates and AD patients associated with impaired Aβ clearance and cerebrovascular accumulation of the peptide.[23,25]

Recent studies have confirmed the role of brain efflux and rapid transport exchanges of Aβ between blood and brain across the BBB in nonhuman primate models of brain parenchymal and cerebrovascular amyloidosis and transgenic models of AD.[25-28] It is of note that we were first to suggest in 1993 that the BBB regulates brain Aβ via specific receptors and/or transporters;[29] the molecular nature of the transport systems and vascular therapeutic targets were not known at that time. It took a decade to characterize at the molecular level different Aβ receptors and transporters at the BBB, define their functions and develop molecular reagents and approaches to confirm the role of transport in pathogenesis of amyloidosis in animal models of AD. After our initial report, a series of papers from different groups have verified the validity of the BBB transport hypothesis for Aβ.[20,23,25-48]

Most recent studies indicate that efflux of Aβ from brain produces rapid increases in plasma Aβ that correlate with the amyloid burden in brain.[25,27] It has been suggested that Aβ efflux measurements may be useful for quantifying brain amyloid burden in patients at risk for or those who have been diagnosed with AD.[27] It has been also demonstrated that development of plaques in nonhuman primate models of amyloidosis AD-type and transgenic animal models of AD shift a transport equilibrium for Aβ between the CNS and plasma due to binding of soluble Aβ from brain and plasma onto amyloid deposits in the CNS and around blood vessels.[25-28,32] Recent studies in squirrel monkeys, a nonhuman primate model of cerebral amyloid angiopathy have demonstrated by single photon emission computed tomography a rapid elimination of Aβ from the brain into plasma suggesting significant clearance of the peptide across the BBB.[25] Studies in primates also indicated an age-related decline in the BBB capacity to eliminate Aβ that correlates with increases in Aβ$_{40/42}$ cerebrovascular immunoreactivity and amyloid deposition.[25,26]

It has been reported that the P-glycoprotein, a member of the ATP-binding cassette (ABC) superfamily of transporters that remove a variety of lipophilic and amphipatic molecules from cells, may eliminate Aβ from brain endothelium via efflux across the apical (blood) facing site

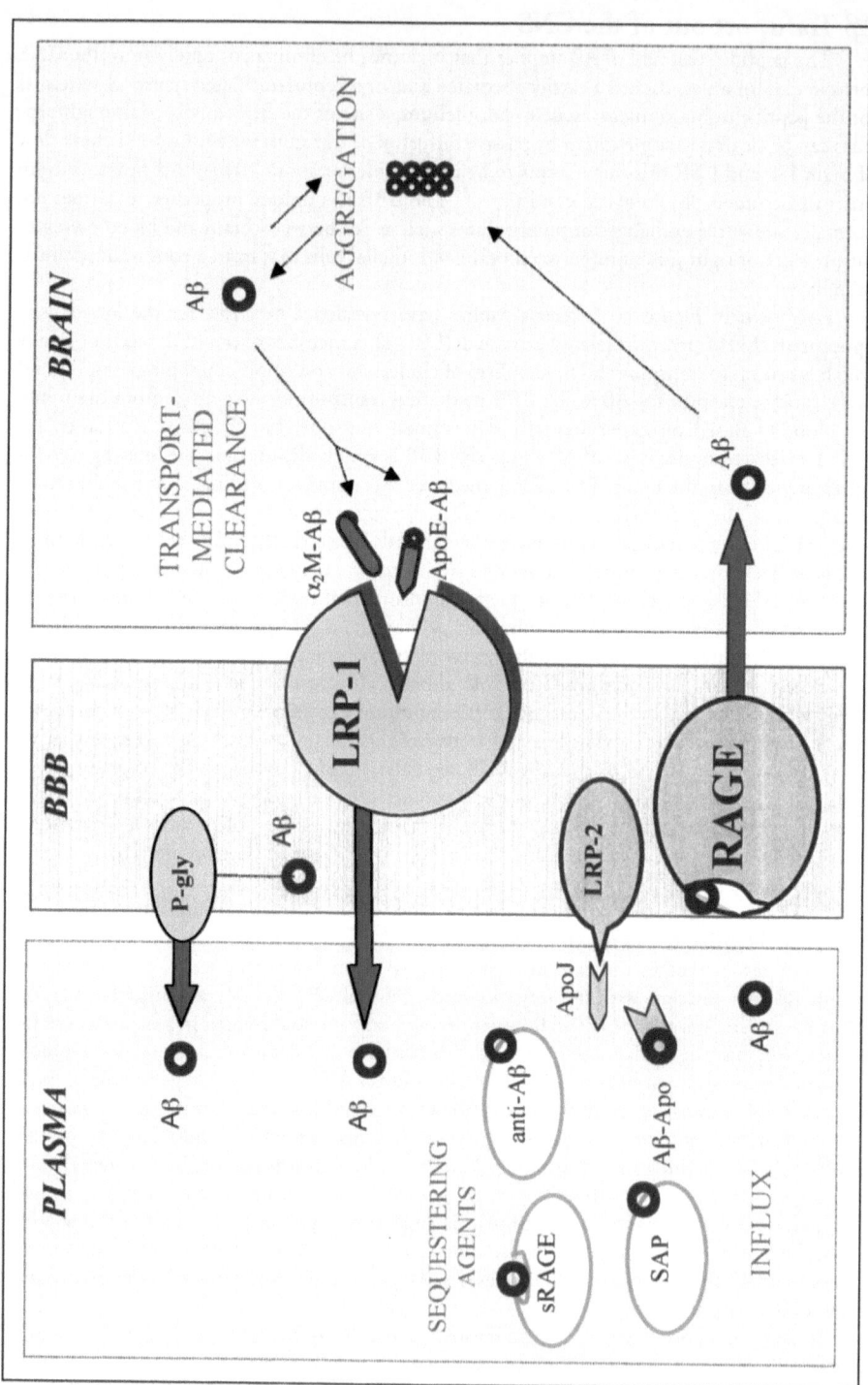

Figure 10.1. Transport-clearance hypothesis for Aβ regulation in the brain (shown on next page). *Transport-mediated clearance of Aβ across the BBB.* LRP-1, a member of the LDL receptor family which is central to metabolism of cholesterol, mediates Aβ transcytosis across the BBB possibly metabolizing the peptide during transport out of the CNS.[23] P-glycoprotein (P-gly), a member of the ABC superfamily of transporters may participate in removing the excess of Aβ from brain endothelium across the luminal site of the BBB.[47] ApoE and α₂-macroglobulin (α₂M), the LRP-1 ligands and transport proteins for Aβ in biological fluids, may influence transport-mediated clearance of Aβ. *Transport of plasma-deived Aβ across the BBB.* LRP-2, a receptor for ApoJ at the luminal site of the BBB, transports ApoJ and ApoJ-Aβ complexes from circulation into the brain.[34] Under physiological conditions LRP-2 is saturated by ApoJ which may preclude transport of Aβ into the CNS. The role of other apolipoproteins (Apo) that bind Aβ in plasma is not presently known. RAGE mediates relatively rapid influx of free, unbound Aβ.[43,54] In AD, RAGE is up-regulated in the vascular system and may act as a "pathogenic" receptor amplifying Aβ accumulation. *Aβ sequestering agents.* Soluble form of the RAGE receptor, sRAGE,[56,57] anti-Aβ IgG[46,61] and/or serum amyloid P component (SAP)[59] can all mop up the Aβ in plasma reducing its influx across the BBB. Eliminating contributions of the circulating pool of Aβ to its central pool could be important in sporadic AD in particular in the presence of defects of the efflux transport systems at the BBB, such as down-regulated LRP-1.[23,25]

of the BBB[47] Thus, the P-glycoprotein cannot directly eliminate Aβ from brain ISF, but may participate in its efflux from brain endothelium into the circulation. The affinity of P-glycoprotein for the peptide is about 500-fold lower compared to LRP-1.

Plasma Aβ Transport

The autosomal dominant mutations that cause early-onset familial AD all increase Aβ (the 42 amino acids isoform) in plasma and brain.[49-53] Recently it has been shown that a novel late onset AD locus on chromosome 10 acts to increase plasma Aβ.[14] At the BBB, LRP-2[34] and the receptor for advanced glycation endproducts (RAGE)[43,54] may transport circulating Aβ into the brain in a complex with apolipoprotein J (ApoJ) or as a free peptide, respectively, which could influence Aβ accumulation. Under physiological conditions LRP-2 is saturated by apoJ which may preclude the entrance of the peptide into the CNS.[34,40] On the other hand, RAGE, a multiligand receptor in the immunoglobulin superfamily, binds soluble Aβ in the nanomolar range and can mediate relatively rapid transport of unbound Aβ into the brain which may result in significant increases in cerebrovascular Aβ if RAGE-mediated influx is not counterbalanced by the efflux from the CNS.[43,54]

In AD, RAGE is up-regulated in the vascular system[55] assuming the role of a "pathogenic" receptor which amplifies Aβ accumulation and mediates vascularly-induced neuronal stress.[54-56] Thus, therapeutic strategies to prevent RAGE-mediated Aβ vascular interactions by either blocking vascular RAGE, or by sequestering circulating peptide with a soluble form of the RAGE receptor, sRAGE, may result in reduced amyloid load, as shown in animal models of systemic amyloidosis and AD.[57,58] Similar, it has been recently reported that removing serum amyloid P component (SAP) from human amyloid deposits in the tissue by drugs that are competitive inhibitors of SAP may provide rapid clearance of SAP, thus producing marked depletion of human circulating SAP, which in turn can provide a new therapeutic approach to both systemic amyloidosis and AD.[59]

Although a key element of Aβ accumulation is decreased clearance of brain amyloid due potentially to decreased exit of Aβ from the brain, there also appears to be a role for intravascular peptide. We propose that there is a complex, but dynamic, equilibrium at least at an early stage in cerebral amyloid accumulation between pathogenic Aβ in the brain and that present in the blood. Trapping Aβ in the intravascular space, especially in the form of complexes with sRAGE, SAP or the antibody (see below), appears to promote its clearance from the CNS.

Immunization Strategies and Transport

Recent studies suggested that the mechanism of action of various immunization or "vaccination" approaches to reduce the amyloid burden in AD transgenic mice,[60,61] and possibly in humans, involves sequestration of plasma-derived Aβ by an anti-Aβ IgG antibody.[46] According to the proposed transport model in Figure 10.1, blockage of Aβ transport into the CNS will not per se increase transport from brain to plasma unless the efflux, i.e., transport out of the CNS, is up-regulated. In AD brains, however, the LRP-1 at the BBB which mediates Aβ transport from brain to blood, is down-regulated.[23] If the anti-Aβ IgG simply mops up the Aβ in plasma the influx will fall and with sustained efflux, the cerebral load of Aβ would decrease as long as the deposited Aβ will resolubilize with time. This implies that the antibody must promote either resolubilization of aggregated Aβ in the brain and/or efflux of a rapidly mobilized soluble pool of Aβ, that should result in Aβ clearing as long the efflux transport systems are intact.

Nonspecific Transport

In addition to the transport systems that rapidly eliminate Aβ across the BBB, a nonspecific bulk flow of ISF can also carry Aβ and/or its metabolites passively from the ISF into the CSF across the permeable ependyma of brain ventricles, and from the CSF back to blood across the arachnoid granulations, choroid plexus and via drainage into deep cervical lymph. According to recent measurements, the ISF-CSF bulk flow has a minor role in physiological clearance of Aβ from the CNS.[23] However, this route may still be important in removing degraded peptide fragments that are normally not recognized by the Aβ putative transporters,[36] and/or under pathological conditions when the excess at Aβ cannot be cleared across the BBB.

Transport-Based Strategies

An independent validation of the transport-clearance hypothesis has been recently obtained with the Dutch mutant peptide, an isoform of Aβ associated with the early onset familial form of amyloidosis AD-type.[48] Substitution at codon 22 resulted in reduced Aβ clearance from the CNS and CSF due to impaired ability of the mutant Aβ peptide to be recognized by the transport systems at the BBB, subsequently leading to its accumulation in the brain.

Only time will confirm whether sporadic AD represents at least, in part, a clearance disorder due to defects in transport of Aβ that is produced normally throughout the life, as suggested by numerous experimental studies. Nevertheless, developing new treatments to lower Aβ by promoting its transport from the CNS should reduce the cellular stress, spontaneous Aβ aggregation, amyloid formation and toxic effects. The body of evidence suggests that transport-based clearance strategies may have potentially important implications for control of cerebral β-amyloidosis in AD, and in conjunction with other Aβ-lowering therapies including immunization/vaccination, resolubilization of amyloid, blockade of plasma Aβ transport and blockade of Aβ production, in eliminating amyloid and plaques from the brain.

Acknowledgement

Our research in AD has been supported by PHS grants AG16223 and NS34467.

References

1. Ghiso J, Frangione B. Amyloidosis in Alzheimer's disease. Adv Drug Del Rev 2002; 54(12):1539-1551.
2. Levy E, Carman MD, Fernandez-Madrid I et al. Mutation of the Alzheimer's disease amyloid gene in hereditary cerebral hemorrhage, Dutch type. Science 1990; 248:1124-1126.
3. Goate A, Chartier-Harlin M-C, Mullan M et al. Segregation of a missense mutation in the amyloid precursor protein gene with familial Alzheimer's disease. Nature 1991; 349:704-706.

4. Schellenberg G. Genetic linkage for a novel familial Alzheimer's disease locus on chromosome 14. Science 1992; 258:868-871.

5. Sherrington R, Rogaev EL, Liang Y et al. Cloning of a gene bearing missense mutations in early onset familial Alzheimer's disease. Nature 1995; 373:754-760.

6. Levy-Lahad E, Wasco W, Poorkaj P et al. Candidate gene for the chromosome 1 familial Alzheimer's disease locus. Science 1995; 269:973-977.

7. Rogaev E, Sherrington R, Rogaeva EA et al. Familial Alzheimer's disease in kindreds with missense mutations in a gene on chromosome 1 related to the Alzheimer's disease type 3 gene. Nature 1995; 376:775-778.

8. Li J, Ma J, Potter, H. Identification and expression analysis of a potential familial Alzheimer disease gene on chromosome 1 related to AD. Proc Natl Acad Sci USA 1995; 92:12180-12184.

9. Corder EH, Saunders AM, Strittmatter WJ et al. Gene dose of apolipoprotein E type 4 allele and the risk of Alzheimer's disease in late onset families. Science 1993; 261:921-923.

10. Strittmatter WJ, Saunders AM, Schmechel D et al. Apolipoprotein E: High-avidity binding to β-amyloid and increased frequency of type 4 allele in late-onset familial Alzheimer disease. Proc Natl Acad Sci USA 1993; 90:1977-1981.

11. Blacker D, Haines JL, Rodes L et al. ApoE-4 and age at onset of Alzheimer's disease: The NIMH genetics initiative. Neurology 1997; 48:139-147.

12. Myers A, Holmans P, Marshall H et al. Susceptibility locus for Alzheimer's disease on chromosome 10. Science 2000; 290:2304-2305.

13. Bertram L, Blacker D, Mullin K et al. Evidence for genetic linkage of Alzheimer's disease to chromosome 10q. Science 2000; 290:2302-2303.

14. Ertekin-Taner N, Graff-Radford N, Younkin LH et al. Linkage of plasma Aβ42 to a quantitative locus on chromosome 10 in late-onset Alzheimer's disease pedigrees. Science 2000; 290:2303-2304.

15. Selkoe DJ Clearing the brain's amyloid cobwebs. Neuron 2001; 32:177-180.

16. Vassar R, Bennett BD, Babu-Khan S et al. Beta-secretase cleavage of Alzheimer's amyloid precursor protein by the transmembrane aspartic protease BACE. Science 1999; 286:735-741.

17. Ray WJ, Yao M, Mumm J et al. Cell surface presenilin-1 participates in the gamma-secretase-like proteolysis of Notch. J Biol Chem 1999; 274:36801-36807.

18. Kamal A, Almenar-Queralt A et al. Kinesin-mediated axonal transport of a membrane compartment containing β-secretase and presenilin-1 requires APP. Nature 2001; 414:643-645.

19. Hardy J, Duff K, Hardy KG et al. Genetic dissection of Alzheimer's disease and related dementias: amyloid and its relationship to tau. Nat Neurosci 1998; 1:355-358.

20. Iwata N, Tsubuki S, Takaki Y et al. Identification of the major Aβ$_{1-42}$-degrading catabolic pathway in brain parenchyma: suppression leads to biochemical and pathological deposition. Nature Med 2000; 6:143-150.

21. Iwata N, Tsubuki S, Takaki Y et al. Metabolic regulation of brain Aβ by neprilysin. Science 2001; 292:1550-1552.

22. Zlokovic BV, Yamada S, Holtzman D et al. Clearance of amyloid β-peptide from brain: transport or metabolism? Nature Med 2000; 6:718-719.

23. Shibata M, Yamada S, Kumar SR et al. Clearance of Alzheimer's amyloid-β$_{1-40}$ peptide from brain by LDL receptor-related protein-1 at the blood-brain barrier. J Clin Invest 2000; 106:1489-1499.

24. Blacker D, Wilcox MA, Laird NM et al. α2-macroglobulin is genetically associated with Alzheimer's disease. Nature Gen 1998; 19:357-360.

25. Bading JR, Yamada S, Mackic JB et al. Brain clearance of Alzheimer's amyloid-β$_{40}$ in the squirrel monkey: a SPECT study in a primate model of cerebral amyloid angiopathy. J Drug Targeting 2002 10; 4:359-368.

26. Mackic JB, Ghiso J, Frangione B et al. Differential cerebrovascular sequestration and enhanced blood-brain barrier permeability to circulating Alzheimer's amyloid-β peptide in aged Rhesus vs. aged Squirrel monkey. Vascular Pharmacology 2002; 18:33-313.

27. DeMattos RB, Bales KR, Cummins DJ et al. Brain to plasma amyloid-β efflux: a measure of brain amyloid burden in a mouse model of Alzheimer's disease. Science 2002; 295:2264-2267.

28. DeMattos RB, Bales KR, Parsadanian M et al. Plaque-associated disruption of CSF and plasma amyloid-β (Aβ) equilibrium in a mouse model of Alzheimer's disease. J Neurochem 2002; 81:229-236J.

29. Zlokovic BV, Ghiso J, Mackic JB et al. Blood-brain barrier transport of circulating Alzheimer's amyloid-β. Biochem Biophys Res Commun 1993; 197:1034-40.

30. Maness LM, Banks WA, Podlisny MB et al. Passage of human amyloid-β protein 1-40 across the murine blood-brain barrier. Life Sci 1994; 55:1643-50.

31. Zlokovic BV, Martel CL, Mackic JB et al. Brain uptake of circulating apolipoproteins J and E complexed to Alzheimer's amyloid-β. Biochem Biophys Res Comm 1994; 205:1431-1437.

32. Ghilardi JR, Catton M, Stimson ER et al. Intra-arterial infusion of $[^{125}I]A\beta_{1-40}$ labels amyloid deposits in the aged primate brain in vivo. Neuroreport 1996; 7:2607-11.

33. Ghersi-Egea JF, Gorevic PD, Ghiso J et al. Fate of cerebrospinal fluid-borne amyloid β-peptide: rapid clearance into blood and appreciable accumulation by cerebral arteries. J Neurochem 1996; 67:880-83.

34. Zlokovic BV, Martel CL, Matsubara E et al. Glycoprotein 330/megalin: probable role in receptor-mediated transport of apolipoprotein J alone and in a complex with Alzheimer's disease amyloid-β at the blood-brain and blood-cerebrospinal fluid barriers. Proc Natl Acad Sci USA 1996; 93:4229-4234.

35. Zlokovic BV. Cerebrovascular transport of Alzheimer's amyloid-β and apolipoproteins J and E: Possible anti-amyloidogenic role of the blood-brain barrier. Life Sci 1996; 59:1483-1497.

36. Banks WA, Kastin AJ. Passage of peptides across the blood brain barrier: pathophysiological perspectives. Life Sci 1996; 59:1923-1943.

37. Martel CL, Mackic JB, McComb JG et al. Blood-brain barrier uptake of the 40 and 42 amino acid sequences of circulating Alzheimer's amyloid-β in guinea pigs. Neurosci.Lett 1996; 206:157-160.

38. Martel CL, Mackic JB, Matsubara E et al. Isoform-specific effects of apolipoproteins E2, E3, E4 on cerebral capillary sequestration and blood-brain barrier transport of circulating Alzheimer's amyloid β. J. Neurochem 1997; 69:1995-2004.

39. Poduslo JF, Curran GL, Haggard JJ et al. Permeability and residual plasma volume of human, Dutch variant, and rat amyloid β-protein 1-40 at the blood-brain barrier. Neurobiol Dis 1997; 4:27-34.

40. Shayo M, McLay RN, Kastin AJ, et al. The putative blood-brain-barrier transporter for the β-amyloid binding protein apolipoprotein J is saturated at physiological concentrations. Life Sci 1997; 60:L115-L118.

41. Zlokovic BV. Can blood-brain barrier play a role in the development of cerebral amyloidosis and Alzheimer's disease pathology. Neurobiol Dis 1997; 4:23-26.

42. Mackic JB, Weiss MH, Miao W et al. Cerebrovascular accumulation and increased blood-brain barrier permeability to circulating Alzheimer's amyloid β-peptide in aged squirrel monkey with cerebral amyloid angiopathy. J. Neurochem 1998 70:210-5.

43. Mackic JB, Stins M, McComb JG et al. Human blood-brain barrier receptors for Alzheimer's amyloid-β. J Clin Invest 1998; 102:734-743.

44. Wegenack TM, Curran GL, Poduslo JF. Targeting Alzheimer amyloid plaques in vivo. Nat Biotechnol 2000; 18:868-872.

45. Poduslo JF, Curran GL. Amyloid-β peptide as a vaccine for Alzheimer's disease involves receptor-mediated transport at the blood-brain barrier. Neuroreport 2001; 12:3197-3200.

46. DeMattos RB, Bales KR, Cummins DJ et al. Peripheral anti-Aβ antibody alters CNS and plasma Aβ clearance and decreases brain Aβ burden in a mouse model of Alzheimer's disease. Proc Natl Acad Sci USA 2001; 98:8850-8855.

47. Lam FC, Liu R, Lu,P et al. β-Amyloid efflux mediated by P-glycoprotein. J Neurochem 2001; 76:1121-1128.

48. Monro OR, Mackic JB, Yamada S et al. Substitution at codon 22 reduces clearance of Alzheimer's amyloid-β peptide from the cerebrospinal fluid and prevents its transport from the central nervous system into blood. Neurobiol Aging 2002; 23:405-412.

49. Citron M, Oltersdorf T, Haase C et al. Mutation of the β-amyloid precursor protein in familial Alzheimer's disease increases β-protein production. Nature 1992; 360:672-674.

50. Cai XD, Golde TE, Younkin SG. Release of excess amyloid-β protein from a mutant amyloid-β_protein precursor. Science 1993; 259:514-516.

51. Suzuki N, Cheung TT, Cai TT et al. An increase percentage of long amyloid-β protein secreted by familial amyloid-β protein precursor (β FF717) mutants. Science 1994; 264:1336-1340.

52. Borchelt DR, Thinakaran G, Eckman CB et al. Familial Alzheimer's disease-linked presenilin 1 variants elevate $A\beta_{1-42/1-40}$ ratio in vitro and in vivo. Neuron 1996; 17:1005-1013.

53. Scheuner D, Eckman C, Jensen M et al. Secreted amyloid-β-protein similar to that in the senile plaques of Alzheimer's disease is increased in vivo by the presenilin 1 and 2 and APP mutations linked to familial Alzheimer's disease. Nature Med 1996; 2:864-870.

54. Kumar R, Miao W, Ghiso J et al. Rage at the blood-brain barrier mediates neurovascular dysfunction caused by amyloid-β_{1-40} peptide. Soc Neurosci Abst 2000; 26:741

55. Yan SD, Chen X, Fu J et al. RAGE and amyloid-beta peptide neurotoxicity in Alzheimer's disease. Nature 1996; 382:685-691.

56. Hofman F, Kumar SR, Maness LM et al. Amyloid-β peptide (1-40) reduction in cerebral blood flow is consequent to RAGE-mediated induction of endothelin-1 Soc Neurosci Abst 2001; 27:333

57. Yan SD, Zhu H, Zhu A et al. Receptor-dependent cell stress and amyloid accumulation in systemic amyloidosis. Nature Med 2000; 6:643-651.

58. Yu J, Zhu H, Pettigrew LC et al. Infusion of soluble RAGE inhibits Aβ amyloid deposition in APP transgenic mice. Soc for Neurosci Abstr 2001; 27:856

59. Pepys MB, Herbert J, Hutchinson GA et al. Targeted pharmacological deletion of serum amyloid P component for treatment of human amyloidosis. Nature 2002; 417: 254-259.

60. Schenk D, Barbour R, Dunn W et al. Immunization with amyloid-β attenuates Alzheimer-disease-like pathology in the PDAPP mouse. Nature 1999; 400:173-177.

61. Sigurdsson EM, Scholtzova H, Mehta PD et al. Immunization with a nontoxic/nonfibrillar amyloid-beta homologous peptide reduces Alzheimer's disease-associated pathology in transgenic mice. Am J Pathol 2001; 159:439-447.

CHAPTER 11

Potential Role of Endogenous and Exogenous Aβ Binding Molecules in Aβ Clearance and Metabolism

Ronald B. DeMattos, Kelly R. Bales, Steven M. Paul and David M. Holtzman

Abstract

Alzheimer's disease (AD) is the leading cause of dementia in the elderly and there are currently no effective therapies for either the prevention or treatment of this disease. The last decade of AD research has been very informative in that major advances have occurred in the understanding of the genetics leading to the early onset familial AD as well as the development of transgenic mouse models which recapitulate several important characteristics of the Alzheimer's pathology. Additionally, substantial efforts have focused on the genesis/synthesis of the Aβ peptide from its precursor protein. A critical area of study that has received less attention is Aβ metabolism. Only until recently have researchers begun to investigate the fate of Aβ following its release from cells into different extracellular compartments. These efforts for the most part have been limited to CNS specific proteolytic events. In this chapter we will review current studies which have begun to dissect the complex systems and molecules that work in concert to regulate Aβ metabolism in both the CNS and peripheral compartments. Unraveling the mechanisms regulating Aβ metabolism in vivo will likely lead to new efforts to improve AD diagnosis as well as rational drug design aimed at both preventing and treating AD.

Introduction

Amyloid-β (Aβ) peptides are predominantly 38-43 amino acids in length and are derived from the amyloid precursor protein (APP) through a series of endoproteolytic cleavages. Abundant evidence over the last 15 years has established that the accumulation in the brain of the normally soluble Aβ peptide into forms with high β-sheet content appears to be central to the pathogenesis of Alzheimer's disease (AD).[1] Genetic and biochemical evidence supporting this idea is that all known mutations that cause early-onset forms of familial AD or Aβ-related cerebral amyloid angiopathy (CAA) map to three genes (APP, presenilin-1 (PS-1), and presenilin-2 (PS-2)). Most of these mutations result in relative overproduction of Aβ$_{42}$, a particularly amyloidogenic form of Aβ, which over time, increases the probability of Aβ aggregation.[1,2] Further, all individuals with Down syndrome possess three copies of APP, have increased levels of Aβ,[3] and all develop AD pathology by age 35.[4]

A major advance in AD research was the development of transgenic mice that over express various human mutated forms of APP genes that lead to familial AD.[5,6] These animals develop age-dependent Aβ accumulation and deposition in their brain. The use of these animal models has been critically important for furthering our understanding of the role of the Aβ peptide in the pathogenesis of AD. Initial studies performed with the transgenic animals yielded confirmatory evidence that paralleled human genetic studies in showing that the enhanced production of the Aβ peptide, more specifically $Aβ_{42}$, led to the pathological deposition of amyloid in brain.[5] Transgenic mice over expressing multiple mutated forms of human APP and either PS1 or PS2 in various combinations result in a gene dose dependent effect on the age of onset of amyloid deposition.[7] These studies clearly demonstrate that significantly increased synthesis of the Aβ peptide will ultimately result in pathology and both amyloid/Aβ dependent and independent memory impairment.[8] However, it must be remembered that the familial mutations in humans that lead to the increased synthesis of Aβ, account for less than 1% of the total cases of AD.

A less investigated area of study in the Aβ-related area of AD has been the in vivo physiological events that occur subsequent to the synthesis of Aβ. Although studying Aβ metabolism post genesis is difficult, it is of paramount importance when considering that >99% of all AD cases result independently from genes known to increase Aβ synthesis. This chapter will focus on recent insights into Aβ metabolism using transgenic animal models of AD. We will review and discuss how Aβ transport into and out of the brain as well as Aβ binding proteins influence Aβ metabolism in vivo.

Animal Models of AD

PDAPP mice, also referred to as APP^{V717F} mice, were the first transgenic model reported to have significant age and region dependent deposition of Aβ.[5] These mice were generated by over expressing a mutated APP minigene (APP^{V717F}) in the brain under control of the platelet derived growth factor (PDGFβ) promoter. In this animal model, the majority of human mutated APP is expressed solely within the central nervous system (CNS).[9] It has been shown that PDAPP animals have similar levels of brain Aβ between birth and ~ 6 months of age; however, they demonstrate marked increases in Aβ levels and Aβ deposition in diffuse and neuritic plaques by 8 months of age in a region specific pattern similar to that seen in AD (deposits in the hippocampus and neocortex but little in striatum and cerebellum)[10] (Fig. 11.1). Also reminiscent of the human condition is the development of neuritic plaques in these APP transgenic mice.[11] Swollen, distorted neuronal processes are associated with the fibrillar Aβ deposits (thioflavine-S positive, i.e., amyloid plaques) (Fig. 11.2). Neuritic dystrophy is not present in areas of diffuse deposits of Aβ. Other features of AD that can be seen in the various APP transgenic mouse models of AD include microglial activation, astrocytosis, evidence of oxidative damage, and some changes in neuronal cytoskeletal proteins including hyperphosphorylation of tau.[12-16] However, neurofibrillary tangles are not seen and little to no neuronal cell death has been observed in any APP transgenic mouse strain, including PDAPP mice. An analysis of the depositing amyloid in aged PDAPP animals shows that the predominant Aβ peptide ends in residue 42 (>90% by 16 months of age).[10] Hsiao and colleagues produced a transgenic mouse model with similar phenotypic findings, Tg2576 or APPsw, wherein a prion protein (PrP) promoter was used to overexpress an APP mutation found in some Swedish families with autosomal dominant AD.[6] Unlike the PDAPP model, these animals express mutated human APP both in the CNS as well as the periphery.[17] In addition to possessing similar brain pathology as observed in PDAPP mice, Tg2576 mice are much more susceptible than PDAPP mice in developing cerebral amyloid angiopathy (CAA).[18] Interestingly, in contrast to PDAPP mice, Tg2576 mice develop Aβ deposits consisting primarily of $Aβ_{1-40}$.[17]

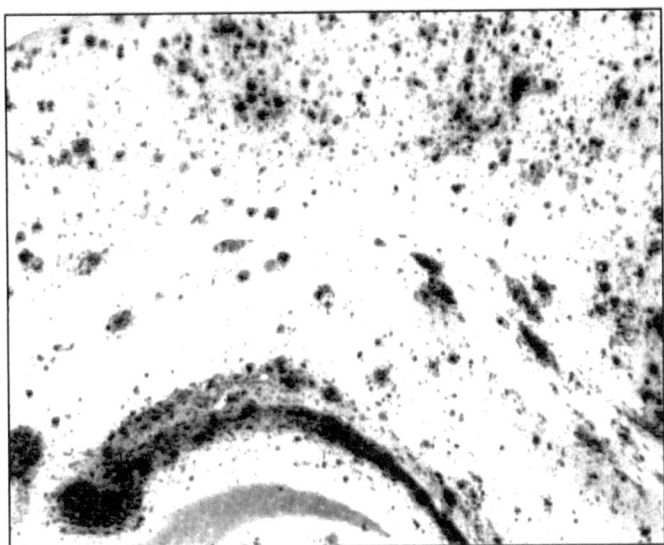

Figure 11.1. Abundant Aβ deposition can be seen in hippocampal and cortical regions in PDAPP$^{+/+}$ transgenic mice. This coronal brain section from a 12-month-old PDAPP mouse was immunostained with a polyclonal antibody against Aβ. Aβ deposition is particularly prominent in the molecular layer of the dentate gyrus as well as in other regions of the hippocampus and cingulate cortex.

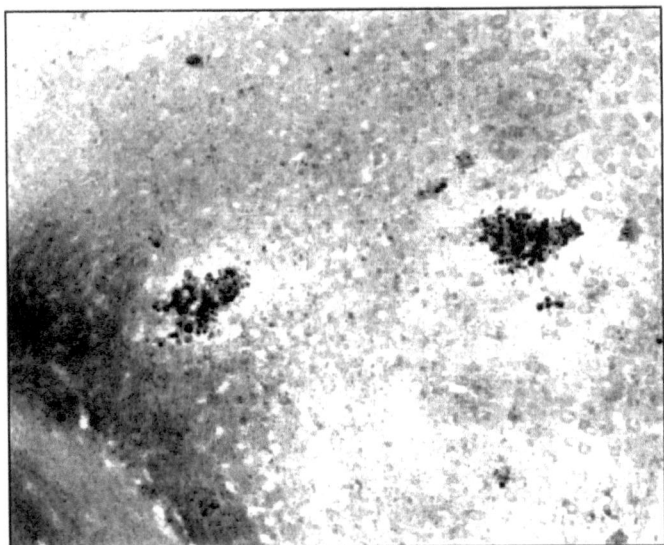

Figure 11.2. PDAPP transgenic mice develop a neuritic dystrophy that is associated with deposited fibrillar Aβ. A coronal brain section from 12-month old PDAPP$^{+/+}$ mouse was stained with the de Olmos silver method to identify the neuritic dystrophy associated with the fibrillar amyloid. Two prominent neuritic plaques are visualized in this section. Each plaque is associated with abundant dystrophic neurites.

Aβ Peptide Metabolism in Human

In vivo metabolism of the Aβ peptide after its initial synthesis is poorly understood due to the complexity of multiple systems that work in concert to facilitate its removal from brain. However, over the past few years several important observations are beginning to yield insight into Aβ's complex metabolism and clearance. Seubert and colleagues were among the first to show that a soluble form of the Aβ peptide could be detected in physiological fluids (CSF and plasma).[19] Subsequent studies revealed that the generation of the Aβ peptide from the parent molecule, APP, occurred in almost all cell types both within the CNS and periphery.[20] Neuronal cells within the CNS robustly express the APP molecule and, as a result, are thought to be the main site for Aβ peptide generation. The exact function of the APP parent molecule remains speculative. Some reports indicate that it may be involved in multiple pathways (proteolytic cascades, axonal transport, neurotropic effects, protein sorting, intracellular signaling, and cellular adhesion), all of which may be nonexclusive.[21-27] Similar to its parent molecule APP, the physiological function of the Aβ peptide remains unknown. Indeed, the hypothesis that the Aβ peptide has a biological function is quite controversial in that many believe it to be a simple byproduct of APP metabolism.

The levels of soluble human Aβ in the CNS are quite high as compared to the periphery (plasma) (Table 11.1). The concentration of human CSF $Aβ_{Total}$ is ~15 ng/ml, whereas plasma $Aβ_{Total}$ levels are much lower at approximately 300 pg/ml.[28,29] The origin of the CSF pool of soluble Aβ is thought to be primarily derived from within the brain. However, the origin of peripheral plasma pool is somewhat speculative in that it is unknown how much is derived from the CNS as opposed to possible peripheral sources. Additionally, the rate of Aβ syntheses in brain is quite robust and is paralleled by rapid clearance mechanisms in both the CNS and peripheral compartments.[30-33] Thus, the concentrations stated above are steady state levels, and represent the highly dynamic processes of synthesis and catabolism. Interestingly, the levels of CSF Aβ appear in some way to be altered by the onset and progression of AD. Several studies have demonstrated that the CSF $Aβ_{42}$ concentrations are significantly decreased in AD patients.[34-36] One possible explanation for the observed decrease in CSF $Aβ_{42}$ is that the deposited amyloid plaque is sequestering it. In apparent contrast to these earlier studies, Jensen and colleagues identified a significant increase in the CSF $Aβ_{42}$ levels early in the progression of the disease and a subsequent significant decrease in CSF $Aβ_{42}$ that positively correlated with the severity of dementia, a finding that the previous studies were unable to identify.[37] It was hypothesized that the reduction of $Aβ_{42}$ in CSF that paralleled the increasing severity of dementia was most likely a result of the decreased number of neuronal cells able to produce Aβ. Our recent studies in human subjects with very mild dementia likely due to AD (also termed mild cognitive impairment) reveals that CSF $Aβ_{42}$ is similar in these subjects vs. age-matched controls (D. Holtzman and A. Fagan, unpublished data). Since the majority of these subjects already have substantial amyloid deposition, the presence of plaques is not associated with a decrease in CSF $Aβ_{42}$ relative to subjects with few to no plaques. Only a few investigations have analyzed the levels of plasma Aβ in AD patients vs. age-matched controls, and there does not appear to be a difference between these groups.[36] Interestingly, one study suggests that cognitively normal, elderly individuals with high levels of plasma Aβ may be more likely to develop dementia over a several year period than those with lower levels.[38] Further studies that analyze human subjects longitudinally both before and after the initiation of the disease will be required to fully understand and characterize the changes in CSF and plasma Aβ and their relationship to AD. These studies are difficult to conduct because the onset and pathological progression of AD is thought to occur 15-20 years prior to the time of onset of even the earliest symptoms of dementia.[39,40]

Table 11.1. Aβ concentrations in CSF and plasma

	Cerebrospinal Fluid	Plasma
	($Aβ_{Total}$)	($Aβ_{Total}$)
Human	14.9 ng/ml[29]	335 pg/ml[28]
PDAPP 3 month old	19.4 ng/ml[46]	176 pg/ml[46]
PDAPP 9 month old	14.1 ng/ml[46]	214 pg/ml[46]
Tg2576 3-6 month old	~56.0 ng/ml[17*]	~19,000 pg/ml[17*]

*The $Aβ_{Total}$ levels shown were derived by the addition of the $Aβ_{40}$ and $Aβ_{42}$ values reported.

Aβ Metabolism in Wild Type and Transgenic Mouse Models of Alzheimer's Disease

Because of the inherent limitations that arise in human studies, researchers have recently begun to utilize wild type mice and APP transgenic mouse models of AD to investigate Aβ metabolism. Initial experiments were designed to recapitulate the genetic findings whereby over expression of the mutant forms of human APP resulted in AD like pathology. Recent studies in transgenic animal models of AD as well as some earlier studies performed in nontransgenic animals have begun to focus on the metabolism of the Aβ peptide prior to the onset of Aβ deposition. Studies first performed by Zlokovic and colleagues showed that exogenously administered ^{125}I-labeled human Aβ to nontransgenic animals could be transported bi-directionally from plasma to CNS and conversely from CNS to plasma.[32,41-43] Additionally, their group demonstrated that the transport was receptor mediated and could be influenced by specific Aβ binding proteins.[32,44] It is unknown, however, to what extent these transport mechanisms modulate the endogenous soluble pools of Aβ between the CNS and periphery. Other investigators have shown that the $Aβ_{40}$ and $Aβ_{42}$ peptides are differentially transported from the CNS to the plasma/periphery. Wisniewski and colleagues demonstrated that $Aβ_{40}$ was rapidly transported from the CNS to plasma with a $t_{1/2}$ of approximately 10 minutes, whereas the CNS to plasma clearance of the $Aβ_{42}$ peptide was much slower showing clearance rates similar to bulk flow of CSF to the periphery.[45] If a significant quantity of endogenously produced CNS Aβ has a similar fate, this metabolic pathway may be a major route of CNS Aβ clearance.

Our group was interested in whether endogenously synthesized human Aβ produced in the CNS is transported to and is in equilibrium with plasma Aβ. For this purpose, we utilized the PDAPP transgenic mouse model where expression of the human APP transgene occurs almost exclusively within the CNS. We analyzed soluble and insoluble steady state Aβ levels in PDAPP transgenic mice that were 3 and 9 months of age.[46] CSF and plasma isolated from young 3-month-old animals had steady state concentrations of Aβ that were very similar to that reported in humans (Table 11.1). Interestingly, there was a positive and highly significant correlation between the concentrations of Aβ in CSF and plasma (r^2 = 0.6392: p < 0.0001). Because the origin of plasma Aβ in PDAPP mice is from the CNS, the data demonstrate that the Aβ in the two compartments (central and peripheral) are in equilibrium. Analysis of a 9-month-old cohort, an age at which Aβ deposition in brain varies from none to heavy, showed that CSF $Aβ_{Total}$ levels were significantly lower (by 28%, Table 11.1, p = 0.01). Additionally, the observed net decrease in the overall levels of soluble Aβ from 3 to 9 months of age was independent of the brain Aβ load. In fact, CSF $Aβ_{Total}$ and $Aβ_{42}$ levels were positively correlated with the amount of Aβ load (Fig. 11.3). Animals that had high levels of Aβ deposition in the cortex had correspondingly high levels of CSF Aβ. Also, the presence of deposited Aβ

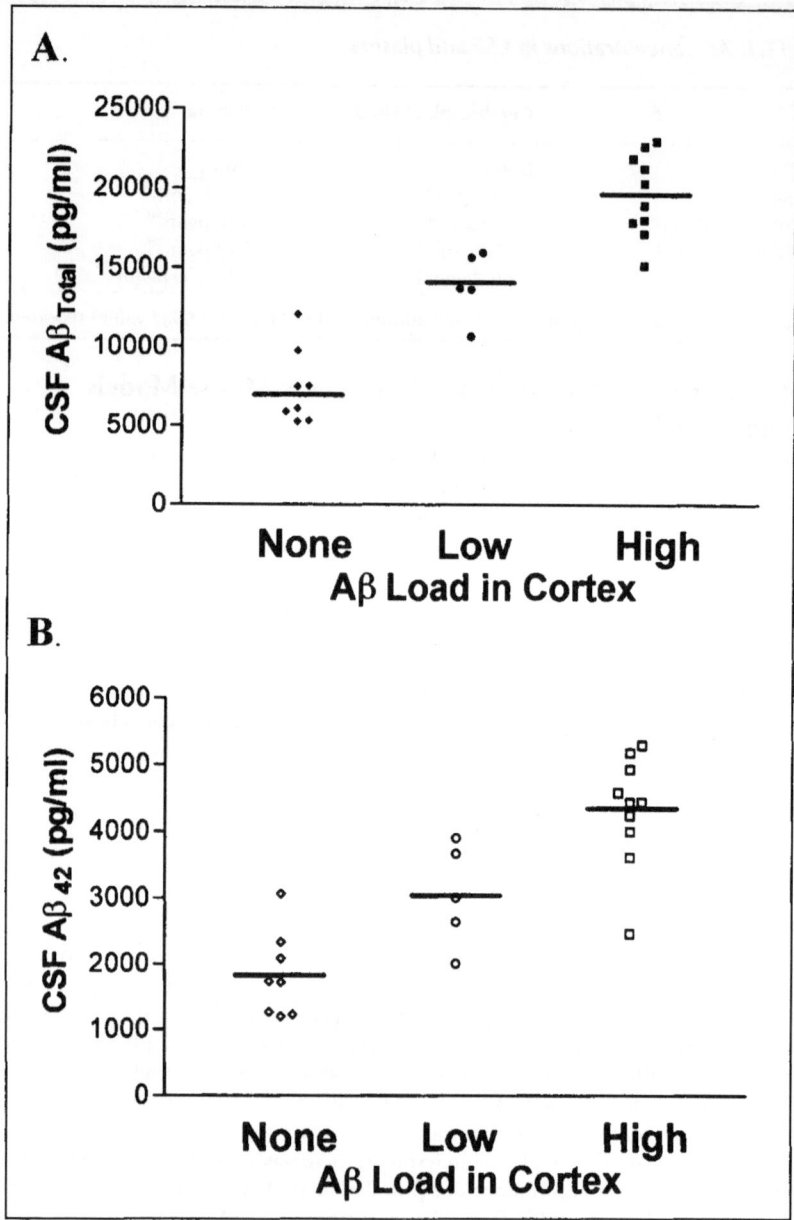

Figure 11.3. CSF $A\beta_{Total}$ and $A\beta_{42}$ correlates with $A\beta$ deposition. CSF was first isolated from 9-month-old PDAPP$^{+/+}$ transgenic mice. Animals were then sacrificed and $A\beta$ load was determined in the cingulate cortex (% of area covered by $A\beta$ immunoreacivity). Strong positive correlations between the amount of cortical $A\beta$ deposition and the levels of CSF $A\beta_{Total}$ ($r^2 = 0.6349$, $p = 0.0001$) and $A\beta_{42}$ ($r^2 = 0.5743$, $p = 0.0001$) were detected. The animal cohort was partitioned into three groups based upon the level of $A\beta$ deposition in the cortex: none, low (<5% $A\beta$ load), and high (>25% $A\beta$ load). CSF $A\beta_{Total}$ (A) and $A\beta_{42}$ (B) levels were significantly increased in a dose dependent manner in animals with increasing levels of $A\beta$ deposition ($p < 0.0001$, ANOVA).

appears to negate the correlation observed between CSF and plasma Aβ in that only animals lacking Aβ deposits maintain a positive correlation between Aβ present in the two compartments. Perhaps there is a parallel between our CSF studies in PDAPP mice with the studies of Jensen et al in humans.[37] They noted a significant increase in Aβ$_{42}$ levels in patients with mild cognitive impairment (MCI) and a subsequent decrease in Aβ levels with the progression of AD which they believe is secondary to neuronal loss/dysfunction. Most patients with MCI already have substantial AD pathology including Aβ amyloid deposition.[39,47] However, neuronal loss remains relatively modest at this stage of disease. Interestingly, PDAPP mice do not develop significant neuronal loss (neurodegeneration) even after high levels of deposited Aβ.[13] The lack of neurodegeneration may unmask the strong relationship between soluble and insoluble Aβ that we have observed (i.e., positive correlation yielding increased levels detected). This interpretation is further bolstered by work from Maggio and colleagues who have shown that even so-called insoluble Aβ can reversibly equilibrate with soluble Aβ.[48]

Kawarabayashi and colleagues also performed a comprehensive study of endogenous Aβ metabolism in a transgenic mouse model of AD.[17] They analyzed age-dependent changes in brain, CSF, and plasma Aβ in Tg2576 mice. As stated above, these animals develop robust amyloid deposition that is similar to that seen in PDAPP mice and in human AD. The CSF and plasma levels of human Aβ in this transgenic animal are quite high as compared to humans (Table 11.1), probably a result of the over expression of APP in both the CNS and periphery. Interestingly, they showed that levels of soluble Aβ in the CSF and plasma decreased significantly during the same time frame of exponential accumulation of brain parenchyma Aβ. Because their analysis focused on groups of animals at various ages as opposed to an individual animal assessment, it is unknown whether the decreased human Aβ levels were solely age related or both age and plaque-related. Furthermore, it may be difficult to discern meaningful relationships between CNS and plasma Aβ in this animal model due predominantly to the fact that most plasma Aβ in these mice is likely derived from other organs and not the CNS. It is also the case that there appears to be more limited variability in Aβ deposition seen in the Tg2576 model at a given age (as compared to the PDAPP model).

Apolipoprotein E and Clusterin: In vivo Aβ Chaperone Proteins

Apolipoprotein E

Aβ binding proteins undoubtedly play an important role in the metabolism of the Aβ peptide. Two of the most widely investigated Aβ binding proteins are apolipoprotein E (ApoE) and clusterin (also known as apolipoprotein J). ApoE is a member of the apolipoprotein family and is primarily known for its role in cholesterol metabolism. ApoE readily associates with plasma lipoproteins (chylomicrons, VLDL, LDL, and HDL) and acts as a mediator of lipoprotein particle uptake via receptor-mediated endocytosis.[49] In the CNS, ApoE is synthesized locally by glial cells (astrocytes and microglia) and is found associated with HDL like particles.[50] Although the role of CNS ApoE is not completely understood, it has been hypothesized that its primary role is for lipid transport.[51,52] There are three common isoforms of ApoE that arise from single amino acid interchanges at position 112 and 158 (ApoE2, ApoE3, and ApoE4). ApoE was first discovered to potentially be an important molecule in Aβ metabolism via genetics. In 1991, researchers at Duke University reported a linkage/association of late-onset AD to a region of chromosome 19. Later studies from this group identified the association of late-onset AD with the epsilon-4 allele of ApoE.[53] Corder et al showed that the risk of AD is increased and the probability of remaining unaffected over time decreases in an ApoE4 dose dependent manner.[54] ApoE4 is also a risk factor for CAA.[55-57] ApoE4 is only partially penetrant and is considered a risk factor for the development of AD since ApoE4 carriers don't invariably develop AD even if they live into their nineties. The fact that ApoE4 is an important AD risk factor for the most common form of AD (late-onset) has led to efforts to identify the pathophysiological role ApoE plays in AD.

Work in the early 1990s demonstrated that ApoE was associated with senile plaques in the brain parenchyma and in amyloid angiopathy.[58,59] Subsequently it was shown that amyloid burden was correlated with ApoE genotype in both AD brains[53] and in patients with traumatic brain injury (increased amyloid deposition in ApoE4-positive subjects).[60] Numerous in vitro biochemical studies demonstrated isoform-specific effects of ApoE on binding to the Aβ peptide, altering Aβ toxicity, and modulating the propensity of the Aβ peptide to aggregate (for review, see refs. 61). Additionally, in vitro studies suggest that ApoE can mediate Aβ clearance by lipoprotein receptor-mediated endocytosis.[62,63] What was unclear, however, from these studies was whether any of these in vitro findings had physiological relevance in vivo.

In 1997 Bales, Paul and colleagues were the first to use a transgenic mouse model of AD to investigate the role of ApoE in the amyloid cascade. They generated PDAPP mice that lacked the expression of murine ApoE (PDAPP, ApoE$^{-/-}$). Analysis of 6-month-old transgenic animals lacking ApoE expression showed a significant decrease in Aβ deposition when compared to transgenic animals expressing murine ApoE. Importantly, these 6-month-old PDAPP, ApoE$^{-/-}$ mice developed only diffuse Aβ deposition and no fibrillar (thioflavine-S or Congo red positive) amyloid plaques. Subsequent studies have shown that PDAPP, ApoE$^{-/-}$ mice have only diffuse plaques until very old ages (e.g., 18 months of age and greater); however, some of these animals eventually develop amyloid plaques.[64] These results suggest that murine ApoE is of critical importance for the conversion of Aβ to a β-sheet conformation that ultimately becomes amyloid. Utilizing a different transgenic mouse model of AD, Tg2576 or APPsw, Holtzman and colleagues confirmed and extended the results described above by demonstrating a similar ApoE dependent phenotype for amyloid deposition as well as a dramatic reduction in the neuritic dystrophy that normally is present in the direct locale of the depositing Aβ.[18] The combined results from the transgenic mouse models lacking endogenous ApoE expression suggest that ApoE is critical for the development of fibrillar amyloid and its associated neuritic dystrophy.

The above studies identified a role of murine ApoE in the amyloid cascade. Our group was interested in whether the human ApoE isoforms would have a similar effect and whether an isoform specific difference between ApoE3 and ApoE4 could be identified. We developed PDAPP transgenic mice expressing human ApoE3 or ApoE4 on the murine ApoE knock out background (PDAPP, ApoE3 or PDAPP, ApoE4).[11,65] The controls in this experiment, PDAPP and PDAPP, ApoE$^{-/-}$ mice, developed abundant Aβ deposits between 9 and 15 months of age. Similar to previous findings, the PDAPP, ApoE$^{-/-}$ mice contained solely diffuse plaques through 15 months (nonfibrillar Aβ deposits) whereas the animals expressing murine ApoE had both diffuse and fibrillar plaques. Interestingly, PDAPP, ApoE3 and PDAPP, ApoE4 mice showed a dramatic suppression of Aβ deposition until ~15 months of age or later. The Aβ that ultimately deposited in these animals (>15 months of age) was both diffuse and fibrillar in nature. Additionally, we observed significantly more Aβ deposition and neuritic dystrophy in ApoE4 versus ApoE3 expressing mice. Recently, we have also found that expression of ApoE2 in PDAPP mice suppressed fibrillar Aβ deposition to an even greater extent than ApoE3.[64] These results demonstrate that human ApoE can isoform-specifically alter the amyloid deposition cascade and the neuritic toxicity associated with amyloid deposition.

Although the mechanism underlying the human ApoE effect on Aβ metabolism remains speculative, the combined results above highlight a few potential mechanisms. A highly probable mechanism through which ApoE may be modulating in vivo Aβ metabolism may be through "clearance". The traditional role of ApoE in the periphery is to act as the ligand for receptor-mediated endocytosis of lipoprotein particles. It is easy to draw a parallel between the function of ApoE in the plasma to that of the CNS, however, the published literature supporting this mechanism is quite limited. In fact, to the authors' knowledge, there is only one manuscript that appears to demonstrate that ApoE-lipoproteins complexes can potentially act as a vehicle for Aβ catabolism.[62] Perhaps a different type of clearance mechanism is at work. Wisniewski et al investigated whether ApoE could modulate the transport of CNS derived Aβ across the blood brain barrier (BBB) into the plasma peripheral system. No significant differences in transport across the BBB were observed for radiolabeled Aβ that was injected intraventricularly in the presence of the various human ApoE isoforms.[45] An important experiment that is currently being conducted in our laboratory is the analysis of all the soluble compartments in PDAPP mice for steady state Aβ levels in the presence of differing levels of human ApoE isoforms. Similar to our characterization of the PDAPP mice at multiple ages (see above), we believe that important insights into ApoE's modulation of Aβ metabolism will be identified and that it is likely that ApoE does modulate Aβ clearance/transport between the CNS and plasma.

Clusterin

The second most abundantly expressed apolipoprotein in the CNS is clusterin (also known as apolipoprotein J).[66-70] CNS clusterin has several similarities to ApoE in that it too is expressed primarily by glia and is present in HDL.[50,71] There are several findings that implicate clusterin as a potential player in Aβ metabolism. Two of the initial observations were that clusterin was upregulated in AD brain and was found associated with deposited amyloid.[72] It was subsequently found that purified clusterin can bind to soluble Aβ$_{40}$ with a dissociation constant characteristic of a high affinity interaction.[73] Other studies have shown that clusterin may be an important regulator of soluble CNS Aβ levels. Studies by Zlokovic et al have shown that Aβ-clusterin complexes can be transported across the blood-brain barrier by a high-affinity receptor mediated process involving transcytosis.[42,44]

Multiple in vitro studies have also highlighted a possible role of clusterin in Aβ metabolism. Several laboratories have shown that clusterin prevents aggregation and polymerization of synthetic Aβ in vitro.[74,75] Additionally, cell culture experiments have demonstrated that Aβ uptake and degradation is facilitated by the presence of clusterin.[76] Oda et al showed that clusterin decreased aggregation of Aβ$_{42}$ and that subsequent incubation of the less aggregated material to PC12 cells significantly increased oxidative stress.[74] Studies by Lambert et al demonstrated that small diffusible oligomers of Aβ$_{42}$ induced by the presence of clusterin were associated with increased neuronal toxicity in organotypic CNS cultures at nanomolar concentrations.[77] Soluble oligomers of Aβ$_{42}$ were also found to be deleterious for LTP in the dentate gyrus of rat hippocampal slices[78] and protofibrillar intermediates of Aβ were found to induce acute electrophysiological and toxic changes to cortical neurons.[79-82] Other recent studies have shown that clusterin can "solubilize" a very broad spectrum of proteins that contain exposed hydrophobic patches.[79,80] Clusterin's chaperone-like activity has been attributed to a molten globule-like region located in the clusterin protein itself.[81,82] While these studies suggest that clusterin/Aβ interactions may be relevant to AD, whether clusterin plays a direct role in the formation of AD pathology in vivo was not clear until recently.

Our group investigated the in vivo role of clusterin in the amyloid cascade by generating PDAPP mice that lacked endogenous clusterin expression (PDAPP, Clu$^{-/-}$).[83] In contrast to the PDAPP, ApoE$^{-/-}$ mice, no significant difference in the relative amount of depositing Aβ was detected between PDAPP or PDAPP, Clu$^{-/-}$ mice. The absence of clusterin did not influence either the age of onset of Aβ deposition or the amount of Aβ accumulation in PDAPP mice. Although there were no differences seen in the quantity of brain Aβ, we did observe significant alterations in the structure of the depositing Aβ (Fig. 11.4). Aβ immunoreactivity in the PDAPP,Clu$^{-/-}$ mice was more diffuse in appearance with fewer "compact" plaques as compared

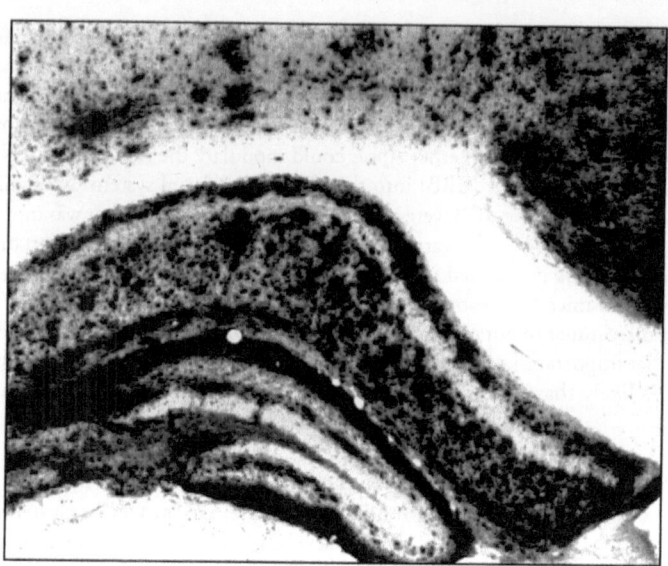

Figure 11.4. Diffuse Aβ deposition can be seen in hippocampal and cortical regions in PDAPP$^{+/+}$,Clu$^{-/-}$ transgenic mice. Coronal brain section from a 12-month-old animal was immunostained with a polyclonal antibody against Aβ. Similar to PDAPP mice expressing clusterin, there was abundant Aβ deposition in the molecular layer of the dentate gyrus. However, there was a more "diffuse" appearance of the Aβ deposits in other regions of the hippocampus and cortex of mice lacking clusterin.

to PDAPP animals expressing endogenous clusterin. Interestingly, the prevalence of this diffuse Aβ deposition was very reminiscent of the type of deposition seen in the PDAPP, ApoE$^{-/-}$ mice. In contrast to the PDAPP, ApoE$^{-/-}$ mice, PDAPP animals lacking clusterin expression did develop thioflavine-S positive Aβ deposits (amyloid); however, there was a significant reduction in the amount and area occupied (% load) by thioflavine-S positive amyloid. Strikingly, the neuritic dystrophy surrounding amyloid deposits in PDAPP, Clu$^{-/-}$ mice was markedly reduced with many deposits having few to no detectable dystrophic neurites. Quantitatively, there was a 10-fold reduction in dystrophic neurites in the hippocampus of PDAPP, Clu$^{-/-}$ versus clusterin expressing mice, and a 5-fold reduction in the number of dystrophic neurites per amyloid deposit. Thus, while the presence or absence of clusterin is associated with amyloid formation, clusterin expression clearly facilitates the neuritic toxicity associated with amyloid.

The dissociation between amyloid formation and neuritic dystrophy in PDAPP, Clu$^{-/-}$ mice implied that clusterin might be influencing a soluble "toxic" species/form of Aβ during or following the process of Aβ deposition. To address this possibility, we assessed the amount of carbonate-soluble brain Aβ by ELISA in cortical brain homogenates under both denaturing and nondenaturing conditions. Our systematic biochemical examination of the carbonate soluble brain extracts with a nondenaturing ELISA specific for oligomeric forms of Aβ detected no significant differences between PDAPP, Clu$^{+/+}$ and PDAPP, Clu$^{-/-}$ mice. However, by using Aβ ELISAs under both denaturing and nondenaturing conditions, we did identify a small but significant two-fold increase in the pool of Aβ which may be monomeric in mice expressing clusterin. Although the exact meaning of this clusterin-dependent alteration of soluble Aβ is unknown, these data provide direct evidence that clusterin modifies Aβ metabolism and (or) structure to influence amyloid deposition and toxicity in vivo.

Exogenous Aβ Binding Proteins

The preceding discussion highlighted the importance of two endogenous Aβ binding proteins and sets the stage for the investigation of the effects of exogenous high affinity Aβ binding proteins on Aβ metabolism in vivo. As was the case with the previous studies, the initial observations were made in vitro. In a two chamber dialysis experiment, it was demonstrated that a high affinity monoclonal antibody directed against the central domain of Aβ, m266, was able to sequester human Aβ from its endogenous CSF binding proteins and create an altered equilibrium favoring Aβ passage across the dialysis membrane (25 kDa cut off).[9] Although this experiment was carried out in a simple dialysis system, the result has important implications.

The data from our dialysis experiment suggested the possibility that an exogenous Aβ binding molecule may be able to alter the concentration gradient and thus the equilibrium of Aβ between the central (brain) and peripheral (blood) compartments in vivo and thereby favor clearance of Aβ from the CNS to the periphery. The latter would promote peripheral Aβ catabolism versus CNS deposition. Our first experiments to test this hypothesis were conducted in young PDAPP animals to see if the m266 antibody would alter the in vivo equilibrium between the CNS and plasma. Intravenous administration of m266 to 3-month-old PDAPP mice resulted in a dramatic accumulation of CNS derived plasma Aβ which was associated with acute changes in CNS steady state levels of Aβ. We demonstrated that all of the plasma Aβ that accumulated after 24 hours following m266 administration (~1000 fold over basal levels) was complexed with antibody. Interestingly, the rate of Aβ entry into the plasma compartment following antibody administration occurred quite rapidly at ~42 pg/ml/min. Part of this massive accumulation appears to be secondary to the antibody decreasing Aβ catabolism once it reaches the plasma from the CNS. We also postulated that part of the increase was due to an increase in the net "flux" of CNS Aβ to the periphery due to an antibody mediated change in Aβ transport between the CNS and plasma compartments. There are at least two mechanisms by which the antibody acting as a "peripheral sink" could alter net Aβ "flux". First, by decreasing the concentration of free (unbound) plasma Aβ to near zero, one would effectively block entry of Aβ from the plasma into the CNS. Second, given a saturable Aβ transporter, such a change in the Aβ concentration gradient across the brain:blood barrier should facilitate efflux until a new equilibrium (with the free plasma Aβ) is reached. Even though our "peripheral sink" hypothesis was supported by data from experiments using exogenous Aβ injections into the CSF space of m266 primed wild type mice, it still remains unclear as to how much of the plasma Aβ accumulating after m266 administration is a result of increased net Aβ CNS efflux or decreased peripheral catabolism of Aβ (i.e., in the presence of the monoclonal antibody). We postulate that both are contributing to the massive plasma Aβ concentrations observed after m266 administration to PDAPP mice.

During the same time course as the massive accumulation of plasma Aβ, we observed a significant increase in CSF Aβ$_{40}$ and Aβ$_{42}$. Although the molecular mechanism underlying the increases in soluble Aβ in CSF is unknown, it may signify an acute alteration in CNS Aβ metabolism due to an exogenously added peripheral Aβ binding protein. Perhaps the shift in equilibrium from the CNS to the periphery alters the concentration dependent Aβ catabolic processes of brain, favoring transport from the parenchyma towards the soluble compartments in CSF and plasma. Importantly, it was shown that this antibody-mediated alteration in Aβ metabolism in PDAPP mice correlated with a significant decrease in plaque pathology following chronic treatment with m266.[9] Bard et al showed similar effects on plaque deposition when using passive administration of different monoclonal or polyclonal antibodies against Aβ.[84] Our studies demonstrated that peripheral administration of an exogenous Aβ binding protein (m266) could alter the in vivo metabolism of Aβ in both the CNS and plasma.

Studies of peripheral administration of m266 to plaque bearing PDAPP mice further support the "flux" hypothesis. As stated previously, CSF Aβ levels positively correlate with the amount of Aβ deposition present in old PDAPP animals. Additionally, plasma Aβ levels in these same mice did not correlate with the extent of Aβ deposition.[46] We wondered whether a single acute injection of m266 into 12 to 13 month old PDAPP animals would detect Aβ entering the periphery from the CNS as well as increase the flux of the soluble CNS pool of Aβ to the periphery. More importantly, we wondered whether the magnitude of this flux would correlate in similar fashion to the amount of deposited brain Aβ. Similar to our previous findings, we showed that plasma levels of Aβ prior to m266 injection did not correlate with the amount of Aβ deposited in brain.[85] However, following acute m266 treatment, striking correlations between plasma Aβ and the amount of deposited Aβ in the brain were observed. In strong support of the flux hypothesis, we demonstrated a highly significant correlation between plasma Aβ and amyloid load as early as 5 minutes post injection of m266 (Fig. 11.5). While inhibition of peripheral catabolism of Aβ by m266 is likely to account for part of the increase in plasma Aβ in all animals, at this 5-minute time point, it is highly unlikely that inhibition of peripheral catabolism could alone account for the correlation between plasma accumulation and plaque burden. In addition to lending support to the flux hypothesis, these experiments also suggest that administering an m266-like antibody to AD patients followed by quantitation of plasma Aβ levels may represent a peripheral biological marker for detecting the presence and quantifying the amount of Aβ-related AD pathology.

Summary

Most studies utilizing transgenic animal models of AD have to date focused on the role of previously identified genes (APP and PS mutations) on Aβ synthesis, deposition, and amyloid plaques. Although these studies are of great importance in verifying the mechanism of some genetic factors involved in AD, they do little to further our knowledge of Aβ metabolism. This chapter focused on the importance of studying Aβ metabolism in vivo. It is imperative that researchers begin to unravel the mechanisms regulating Aβ metabolism in addition to Aβ synthesis. This will likely lead to new efforts in rational drug design aimed at preventing and treating AD. In our own laboratories we have analyzed both the soluble and insoluble pools of Aβ before and after Aβ deposition has occurred. Observed correlations between CNS and plasma pools of Aβ indicate a system of communication between the central and peripheral compartments and thereby demonstrate ordered processes which regulate the clearance of the Aβ peptides from the brain into the circulation followed by its removal. Interestingly, studies performed with the monoclonal antibody m266 demonstrated that the rate of CNS derived Aβ entry into the plasma is very rapid (~ 42 pg/ml/min) and implies that the peripheral compartment is normally a major site for the catabolism of Aβ (steady state levels are maintained between 200 to 500 pg/ml). A better understanding of the pathways by which Aβ is transported between the CNS and periphery either across the blood:brain barrier as well as via interstitial fluid and CSF bulk flow pathways could provide major insights into why Aβ is deposited and amyloid formation occurs in AD and CAA. Studies performed in vivo have also shown that ApoE and clusterin are important endogenous Aβ binding proteins that can modulate both the final form (fibrillar or amorphous), level, and associated toxicity of the depositing Aβ peptide. However, the molecular mechanisms underlying these effects remain unknown. Based upon the human genetics and supporting transgenic mouse studies, it is likely that the ApoE isoform-specific delay in deposition is being manifested through alterations in Aβ clearance/metabolism. We postulate that a careful inspection and understanding of Aβ metabolism in human ApoE transgenic animals will also yield novel insights and new therapeutic opportunities for intervention.

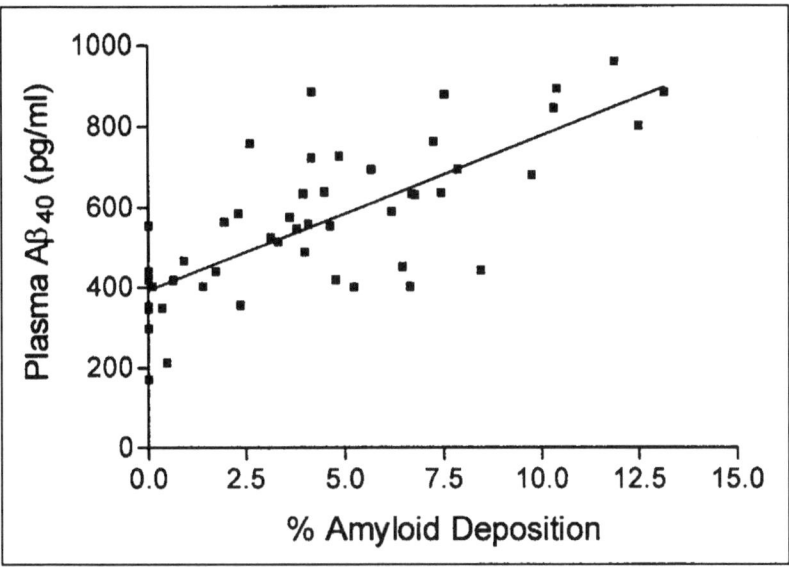

Figure 11.5. Plasma levels of Aβ40 following m266 administration are highly correlated with Aβ and amyloid burden in hippocampus in 12-13 month old PDAPP mice. Before and 5 minutes after the i.v. administration of m266 (500 μg), plasma samples were collected. Prior to m266 administration, there was no correlation between the plasma levels of Aβ40 and amyloid load in the hippocampus. In contrast, 5 minutes following i.v. administration of m266, there was a highly significant correlation between plasma levels of Aβ40 and the amyloid load in the hippocampus (r = 0.7420, p< 0.0001).

References

1. Selkoe DJ. Alzheimer's disease: genes, proteins, and therapy. Physiol Rev 2001; 81:741-66.
2. Sisodia SS. Alzheimer's disease: perspectives for the new millennium. J Clin Invest 1999; 104:1169-70.
3. Rumble B, Retalack R, Hilbich C et al. Amyloid A4 protein and its precursors in Down's syndrome and Alzheimer's disease. N Engl J Med 1989; 320:1446-52.
4. Wisniewski KE, Wisniewski HM, Wen GY. Occurrence of neuropathological changes and dementia of Alzheimer's disease in Down's syndrome. Ann Neurol 1985; 17:278-82.
5. Games D, Adams D, Alessandrini R et al. Alzheimer-type neuropathology in transgenic mice overexpressing V717F β-amyloid precursor protein. Nature 1995; 373:523-27.
6. Hsiao K, Chapman P, Nilsen S et al. Correlative memory deficits, Aβ elevation, and amyloid plaques in transgenic mice. Science 1996; 274:99-102.
7. Guenette SY, Tanzi RE. Progress toward valid transgenic mouse models for Alzheimer's disease. Neurobiol. Aging 1999; 20:201-11.
8. Dodart J-C, Mathis C, Bales KR et al. Does my mouse have Alzheimer's disease? Genes, Brain and Behavior 2002; 1:142-55.
9. DeMattos RB, Bales KR, Cummins DJ et al. Peripheral anti-Aβ antibody alters CNS and plasma Aβ clearance and decreases brain Aβ burden in a mouse model of Alzheimer's disease. Proc Natl Acad Sci USA 2001; 98:8850-55.
10. Johnson-Wood K, Lee M, Motter R et al. Amyloid precursor protein processing and Aβ42 deposition in a transgenic mouse model of Alzheimer's disease. Proc Natl Acad Sci USA 1997; 94:1550-55.
11. Holtzman DM, Bales KR, Tenkova T et al. Apolipoprotein E isoform-dependent amyloid deposition and neuritic degeneration in a mouse model of Alzheimer's disease. Proc Natl Acad Sci USA 2000; 97:2892-97.

12. Irizarry MC, McNamara M, Fedorchak K et al. APPsw transgenic mice develop age-related Aβ deposits and neuropil abnormalities, but no neuronal loss in CA1. J Neuropath Exp Neurol 1997; 56:965-73.

13. Irizarry MC, Soriano F, McNamara M et al. Aβ deposition is associated with neuropil changes, but not with overt neuronal loss in the human amyloid precursor protein V717F (PDAPP) transgenic mouse. J Neurosci 1997; 17:7053-59.

14. Masliah E, Sisk A, Mallory M et al. Comparison of neurodegenerative pathology in transgenic mice overexpressing V717F β-amyloid precursor protein and Alzheimer's disease. J Neurosci 1996; 16:5795-811.

15. Smith MA, Hirai K, Hsiao K et al. Amyloid-β deposition in Alzheimer transgenic mice is associated with oxidative stress. J Neurochem 1998; 70:2212-15.

16. Frautschy SA, Yang F, Irizarry M et al. Microglial response to amyloid plaques in APPsw transgenic mice. Am J Pathol 1998; 152:307-17.

17. Kawarabayashi T, Younkin LH, Saido TC et al. Age-dependent changes in brain, CSF, and plasma amyloid β protein in the Tg2576 transgenic mouse model of Alzheimer's disease. J Neurosci 2001; 21:372-81.

18. Holtzman DM, Fagan AM, Mackey B et al. ApoE facilitates neuritic and cerebrovascular plaque formation in the APPsw mouse model of Alzheimer's disease. Ann Neurol 2000; 47:739-47.

19. Seubert P, Vigo-Pelfrey C, Esch F et al. Isolation and quantification of soluble Alzheimer's β-peptide from biological fluids. Nature 1992; 359:325-27.

20. Selkoe DJ. Amyloid beta-protein and the genetics of Alzheimer's disease. J Biol Chem 1996; 271:18295-98.

21. Sinha S, Dovey HF, Seubert P et al. The protease inhibitory properties of the Alzheimer's beta-amyloid precursor protein. J Biol Chem 1990; 265:8983-85.

22. Smith RP, Higuchi DA, Broze GJJ. Platelet coagulation factor XIa-inhibitor, a form of Alzheimer amyloid precursor protein. Science 1990; 248:1126-28.

23. Van Nostrand WE, Schmaier AH, Farrow JS et al. Protease nexin-II (amyloid beta-protein precursor): a platelet alpha-granule protein. Science 1990; 248:745-48.

24. Saitoh T, Sundsmo M, Roch JM et al. Secreted form of amyloid beta protein precursor is involved in the growth regulation of fibroblasts. Cell 1989; 58:615-22.

25. Schubert D, Jin LW, Saitoh T et al. The regulation of amyloid beta protein precursor secretion and its modulatory role in cell adhesion. Neuron 1989; 3:689-94.

26. Small DH, Nurcombe V, Reed G et al. A heparin-binding domain in the amyloid protein precursor of Alzheimer's disease is involved in the regulation of neurite outgrowth. J Neurosci 1994; 14:2117-27.

27. Kamal A, Almenar-Queralt A, LeBlanc JF et al. Kinesin-mediated axonal transport of a membrane compartment containing beta-secretase and presenilin-1 requires APP. Nature 2001; 414:643-48.

28. Lannfelt L, Basun H, Vigo-Pelfrey C et al. Amyloid beta-protein in cerebrospinal fluid in individuals with the swedish Alzheimer amyloid precursor protein mutation. Neuroscience Letters 1995; 199:203-06.

29. Matsubara E, Ghiso J, Frangione B et al. Lipoprotein-free amyloidogenic peptides in plasma are elevated in patients with sporadic Alzheimer's disease and Down's syndrome. Ann Neurol 1999; 45:537-41.

30. Savage MJ, Trusko SP, Howland DS et al. Turnover of amyloid beta-protein in mouse brain and acute reduction of its level by phorbol ester. J Neurosci 1998; 18:1743-52.

31. Saido TC. Alzheimer's disease as proteolytic disorders: anabolism and catabolism of beta-amyloid. Neurobiology of Aging 1998; 19:S69-S75.

32. Shibata M, Yamada S, Kumar SR et al. Clearance of Alzheimer's amyloid-β_{1-40} peptide from brain by LDL receptor-related protein-1 at the blood-brain barrier. J Clin Invest 2000; 106:1489-99.

33. Iwata N, Tsubuki S, Takaki Y et al. Metabolic regulation of brain Amyloid-beta by neprilysin. Science 2001; 292:1550-52.

34. Motter R, Vigopelfrey C, Kholodenko D et al. Reduction of beta-amyloid peptide (42) in the cerebrospinal fluid of patients with Alzheimer's disease. Ann Neurol 1995; 38:643-48.

35. Galasko D, Chang L, Motter R et al. High cerebrospinal fluid tau and low amyloid β42 levels in the clinical diagnosis of Alzheimer disease and relation to apolipoprotein E genotype. Arch Neurol 1998; 55:937-45.
36. Mehta PD, Pirttila T, Mehta SP et al. Plasma and cerebrospinal fluid levels of amyloid β proteins 1-40 and 1-42 in Alzheimer's disease. Arch Neurol 2000; 57:100-05.
37. Jensen M, Schroder J, Blomber M et al. Cerebrospinal fluid Aβ42 is increased early in sporadic Alzheimer's disease and declines with disease progression. Ann Neurol 1999; 45:504-11.
38. Mayeux R, Tang MX, Jacobs DM et al. Plasma amyloid β-peptide 1-42 and incipient Alzheimer's disease. Ann Neurol 1999; 46:412-16.
39. Price JL, Morris JC. Tangles and plaques in nondemented aging and "preclinical" Alzheimer's disease. Ann Neurol 1999; 45:358-68.
40. Yamaguchi H, Sugihara S, Ogawa A et al. Alzheimer β amyloid deposition enhanced by ApoE epsilon 4 gene precedes neurofibrillary pathology in the frontal association cortex of nondemented senior subjects. J Neuropathol Exp Neurol 2001; 60:731-39.
41. Zlokovic BV, Ghiso J, Mackic JB et al. Blood-brain barrier transport of circulating Alzheimer's amyloid β. Biochem Biophys Res Comm 1993; 197:1034-40.
42. Zlokovic BV, Martel CL, Mackic JB, Matsubara E et al. Brain uptake of circulating apolipoproteins J and E complexed to Alzheimer's amyloid β. Biochem Biophys Res Commun 1994; 205:1431-37.
43. Ghersi-Egea J-F, Gorevic PD, Ghiso J et al. Fate of cerebrospinal fluid-borne amyloid β -peptide: Rapid clearance into blood and appreciable accumulation by cerebral arteries. J Neurochem 1996; 67:880-83.
44. Zlokovic BV, Martel CL, Matsubara E et al. Glycoprotein 330/megalin: probable role in receptor-mediated transport of apolipoprotein J alone and in a complex with Alzheimer's disease amyloid β at the blood-brain and blood-cerebrospinal fluid barriers. Proc Natl Acad Sci USA 1996; 93:4229-34.
45. Ji Y, Permanne B, Sigurdsson EM, et al. Amyloid β$_{40/42}$ clearance across the blood-brain barrier following intraventricular injections in wild-type, ApoE knock-out and human ApoE3 or E4 expressing transgenic mice. J Alzheimer's Dis 2001; 3:23-30.
46. DeMattos RB, Bales KR, Parsadanian M et al. Plaque-associated disruption of CSF and plasma Aβ equilibrium in a mouse model of Alzheimer's disease. J Neurochem 2002; 81:229-36.
47. Morris JC, Storandt M, Miller JP et al. Mild cognitive impairment represents early-stage Alzheimer disease. Arch Neurol 2001; 58:397-405.
48. Esler WP, Stimson ER, Jennings JM et al. Alzheimer's disease amyloid propagation by a template-dependent dock-lock mechanism. Biochemistry 2000; 39:6288-95.
49. Mahley RW. Apolipoprotein E: Cholesterol transport protein with expanding role in cell biology. Science 1988; 240:622-30.
50. DeMattos RB, Brendza RP, Heuser JE et al. Purification and characterization of astrocyte-secreted apolipoprotein E and J-containing lipoproteins from wild-type and human ApoE transgenic mice. Neurochem Int 2001; 39:415-25.
51. Poirer J. Apolipoprotein E in animal models of CNS injury and Alzheimer's disease. Trends Neurol Sci 1994; 17:525-30.
52. Holtzman DM, Fagan AM. Potential role of ApoE in structural plasticity in the nervous system: Implications for diseases of the central nervous system. Trends Cardiovasc. Med 1998; 8:250-55.
53. Strittmatter WJ, Saunders AM, Schmechel D et al. Apolipoprotein E: high avidity binding to β-amyloid and increased frequency of type 4 allele in late-onset familial Alzheimer disease. Proc Natl Acad Sci USA 1993; 90:1977-81.
54. Corder EH, Saunders AM, Strittmatter WJ et al. Gene dose of apolipoprotein E type 4 allele and the risk of Alzheimer's disease in late onset families. Science 1993; 261:921-23.
55. Schmechel DE, Saunders AM, Strittmattter WJ et al. Increased amyloid β-peptide deposition in cerebral cortex as a consequence of apolipoprotein genotype in late-onset Alzheimer disease. Proc Natl Acad Sci USA 1993; 90:9649-53.
56. Greenberg SM, Rebeck GW, Vonsattel JPG et al. Apolipoprotein E e4 and cerebral hemorrhage associated with amyloid angiopathy. Ann Neurol 1995; 38:254-59.
57. Greenberg SM, Briggs ME, Hyman BT et al. Apolipoprotein E e4 is associated with the presence and earlier onset of hemorrhage in cerebral amyloid angiopathy. Stroke 1996; 27:1333-37.

58. Namba Y, Tomonaga M, Kawasaki et al. Apolipoprotein E immunoreactivity in cerebral amyloid deposits and neurofibrillary tangles in Alzheimer's disease kuru plaque amyloid in Creutzfeldt-Jacob disease. Brain Res 1991; 541:163-66.
59. Wisniewski T, Frangione B. Apolipoprotein E: a pathological chaperone protein in patients with cerebral and systemic amyloid. Neurosci Lett 1992; 135:235-38.
60. Nicoll JAR, Roberts GW, Graham DI. ApoE E4 allele is associated with deposition of amyloid beta-protein following head injury. Nature Med 1995; 1:135-37.
61. Holtzman DM. Role of ApoE/Aβ interactions in the pathogenesis of Alzheimer's disease and cerebral amyloid angiopathy. J Mol Neurosci 2001; 17:147-55.
62. Urmoneit B, Prikulis I, Wihl G et al. Cerebrovascular smooth muscle cells internalize Alzheimer amyloid beta protein via a lipoprotein pathway: implications for cerebral amyloid angiopathy. Lab Invest 1997; 77:157-66.
63. Yang DS, Small DH, Seydel U et al. Apolipoprotein E promotes the binding and uptake of beta-amyloid into Chinese hamster ovary cells in an isoform-specific manner. Neurosci 1999; 90:1217-26.
64. Fagan AM, Watson M, Parsadanian M et al. Human and murine apoE markedly influence Aβ metabolism both prior and subsequent to plaque formation in a mouse model of Alzheimer's disease. Neurobiol Dis 2002; 9:305-318.
65. Holtzman DM, Bales KR, Wu S et al. In vivo expression of apolipoprotein E reduces amyloid-β deposition in a mouse model of Alzheimers disease. J Clin Invest 1999; 103:R15-R21.
66. Roheim PS, Carey M, Forte T et al. Apolipoproteins in human cerebrospinal fluid. Proc Natl Acad Sci USA 1979; 76:4646-49.
67. Hochstrasser A-C, James RW, Martin BM et al. HDL particle associated proteins in plasma and cerebrospinal fluid: identification and partial sequencing. Appl Theoret Electroph 1988; 1:73-76.
68. May PC, Finch CE. Sulfated glycoprotein 2: new relationships of this multifunctional protein to neurodegeneration. Trends Neurol Sci 1992; 15:391-96.
69. Aronow BJ, Lund SD, Brown TL et al. Apolipoprotein J expression at fluid-tissue interfaces: Potential role in barrier cytoprotection. Proc Natl Acad Sci USA 1993; 90:725-29.
70. Borghini I, Barja F, Pometta D et al. Characterization of subpopulations of lipoprotein particles isolated from human cerebrospinal fluid. Biochem Biophys Acta 1995; 1255:192-200.
71. LaDu MJ, Gilligan SM, Lukens SR et al. Nascent astrocyte particles differ from lipoproteins in CSF. J Neurochem 1998; 70:2070-81.
72. May PC, Lampert-Etchells M, Johnson SA et al. Dynamics of gene expression for a hippocampal glycoprotein elevated in Alzheimer's disease and in response to experimental lesions in rat. Neuron 1990; 5:831-39.
73. Matsubara E, Frangione B, Ghiso J. Characterization of apolipoprotein J-Alzheimer's Aβ interaction. J Biol Chem 1995; 270:7563-67.
74. Oda T, Wals P, Osterburg HH et al. Clusterin (ApoJ) alters the aggregation of amyloid β-peptide (Aβ1-42) and forms slowly sedimenting Aβ complexes that cause oxidative stress. Exp Neurol 1995; 136:22-31.
75. Matsubara E, Soto C, Governale S et al. Apolipoprotein J and Alzheimer's amyloid beta solubility. Biochem J 1996; 316:671-79.
76. Hammad SM, Ranganathan S, Loukinova E et al. Interaction of apolipoprotein J-amyloid beta-peptide complex with low density lipoprotein receptor-related protein-2/megalin. A mechanisms to prevent pathological accumulation of amyloid beta peptide. J Biol Chem 1997; 272:18644-49.
77. Lambert MP, Barlow AK, Chromy BA et al. Diffusible, nonfibrillar ligands derived from Aβ1-42 are potent central nervous system neurotoxins. Proc Natl Acad Sci USA 1998; 95:6448-53.
78. Wang H-W, Pasternak JF, Kuo H et al. Soluble oligomers of beta amyloid (1-42) inhibit long-term potentiation but not long-term depression in rat dentate gyrus. Brain Research 2002; 924:133-140.
79. Humphreys DT, Carver JA, Easterbrook-Smith SB et al. Clusterin has chaperone-like activity similar to that of small heat shock proteins. J Biol Chem 1999; 274:6875-81.
80. Poon S, Easterbrook-Smith SB, Rybchyn MS et al. Clusterin is an ATP-independent chaperone with very broad substrate specificity that stabilizes stressed proteins in a folding-competent state. Biochemistry 2000; 39:15953-60.

81. Bailey RW, Dunker AK, Brown CJ et als. Apolipoprotein J expression at fluid-tissue interfaces: Potential role in barrier cytoprotection. Proc Natl Acad Sci USA 2001; 90:725-29.

82. Dunker AK, Lawson JD, Brown CJ et al. Intrinsically disordered protein. Journal of Molecular Graphics and Modelling 2001; 19:26-59.

83. DeMattos RB, O'Dell MA, Parsadanian M et al. Clusterin promotes amyloid plaque formation and is critical for neuritic toxicity in a mouse model of Alzheimer's disease. Proc Natl Acad Sci 2002; 99:10843-48.

84. Bard F, Cannon C, Barbour R et al. Peripherally administered antibodies against amyloid β-peptide enter the central nervous system and reduce pathology in a mouse model of Alzheimer's disease. Nature Med 2000; 6:916-19.

85. DeMattos RB, Bales K, Cummins DJ et al. Brain to plasma amyloid-β efflux: A measure of brain amyloid burden in a mouse model of Alzheimer's disease. Science 2002; 295:2264-67.

Modulating Amyloid-β Levels by Immunotherapy:

A Potential Therapeutic Strategy for the Prevention and Treatment of Alzheimer's Disease

Cynthia A. Lemere, Timothy J. Seabrook, Melitza Iglesias, Chica Mori, Jodi F. Leverone and Edward T. Spooner

Abstract

Alzheimer's disease (AD) is the most common form of dementia and afflicts ~15-20 million people worldwide. Currently, there is no effective cure. Research efforts over the past decade have demonstrated that amyloid-beta protein (Aβ), a small peptide generated from its large precursor protein, the amyloid precursor protein (APP), plays a central role in AD pathogenesis, thus leading to the development of therapies aimed at lowering Aβ levels in the brain. One such strategy involves using Aβ immunotherapy, either by direct Aβ peptide vaccination or passive transfer of Aβ-specific antibodies, to modulate Aβ levels in the central nervous system (CNS). Here, we provide an overview of such immune-based studies in wildtype (WT) mice and transgenic (tg) mouse models of AD, as well as those in humans. Aβ immunization in APP tg mice has proven effective in lowering cerebral Aβ levels, including plaque deposition, with the caveat that the earlier it is given, the better. Prevention of and improvement in behavior deficits normally seen in APP tg mice has been shown following both active and passive Aβ immunization. Various immunization protocols have been tested in both WT and tg mice and are described here. Three proposed mechanisms for the Aβ-lowering, behavioral improvement effects of the Aβ vaccine are discussed and include: disaggregation of Aβ fibrillar aggregates and prevention of soluble Aβ to form fibrils, Fc-mediated microglial phagocytosis of Aβ, and a shift in efflux of Aβ from CNS to the periphery. Following clinical safety trials, a large Phase IIa clinical Aβ vaccine trial in AD patients was initiated in the USA and Europe in late 2001. The dosing was stopped on March 1, 2002, due to adverse CNS reactions in ~5% of patients. We speculate here about the possible causes for such adverse effects and provide a culmination of ideas from many researchers towards the future of Aβ immunotherapy for the prevention and treatment of AD. It is with optimism that we proceed.

Aβ Immunotherapy Prevents or Reduces AD Pathology and Improves Behavioral Deficits in Transgenic Mouse Models of AD

In 1999, Schenk et al showed for the first time that immunizing young PDAPP tg mice (bearing mutant human APP/V717F) prior to Aβ deposition essentially prevented cerebral Aβ plaque formation, while immunizing after Aβ deposition had begun, led to a decrease in plaque burden.[1] Monthly intraperitoneal (i.p.) injections of aggregated $Aβ_{1-42}$ were given to mice from either 6 weeks to 13 months of age or from 11 to 18 months of age. In the younger mice, Aβ immunization not only prevented plaque formation, but it also prevented the gliosis and neuritic dystrophy that accompanies Aβ plaque formation in APP tg mouse brain. In the older mice, immunization suppressed new plaque formation and may have removed some of the existing plaques. At 18 months, the remaining plaques were mostly compacted and were often associated with MHC class II-immunoreactive (IR) activated microglia.

Since the publication of Schenk's findings, numerous reports have confirmed and extended these original results. For example, in a subsequent report from the same group, Bard et al, demonstrated that passive transfer of certain Aβ-specific monoclonals, 3D6 and 10D5, and a polyclonal against $Aβ_{1-42}$, via weekly i.p. injections for 6 months also reduced brain levels of Aβ in PDAPP mice.[2] Not all Aβ antibodies tested were efficacious in lowering cerebral Aβ. Their data show that the Aβ antibodies, themselves, had a direct effect on modulating Aβ in the brain, thus eliminating the dependence upon a cellular immune response. In the same year, and in collaboration with Drs. Weiner and Selkoe, we reported that weekly intranasal (i.n.) administration of $Aβ_{1-40}$ to PDAPP tg mice from 5 to 12 months of age resulted in the generation of low titers of anti-Aβ antibodies (~26 μg/ml) and a significant 56% reduction in cerebral Aβ burden (quantified biochemically and immunohistochemically).[3,4] In 2001, Vehmas et al showed that monthly i.p. immunization with $Aβ_{1-42}$ in another AD mouse model with accelerated pathology, PSAPP double transgenic mice [bearing familial mutations in both human APP and presenilin 1 (PSI)], also resulted in the generation of anti-Aβ titers, a significant reduction in Aβ deposits in brain and an increase in cerebral levels of soluble Aβ.[5] Many of the Aβ immunization studies showing decreased cerebral Aβ burden were conducted in relatively young mice either prior to plaque deposition or in its early stages. Das et al tested Aβ immunization by monthy i.p. injection of $Aβ_{1-42}$ at three different time points in Tg2576 APP tg mice and found that while the generation of antibodies prevented or slowed plaque deposition if given early enough, there was not much of a change in cerebral Aβ levels in old mice vaccinated after abundant compacted plaques had become deposited.[6]

Behavioral studies have demonstrated that one of the beneficial effects of Aβ immunization in AD tg mouse models is behavioral improvement. In December, 2000, Janus et al reported that in another APP tg mouse model, TgCRND8, five i.p. injections of $Aβ_{1-42}$ between 6 and 16 weeks of age led to reduced plaque burden and improved behavior in a reference memory version of the Morris water maze test.[7] At the same time, Morgan et al reported that in two additional AD tg mouse models (Tg2576 and PSAPP) chronic subcutaneous (s.c.) injection of $Aβ_{1-42}$ from 7.5 to 15.5 months of age resulted in improved behavior in a radial-arm maze test of working memory and only a modest reduction in plaque burden.[8] Subsequently, the same group reported that the beneficial effects of Aβ immunization on behavior in AD mouse models may be task-specific. Arendash et al showed that chronic s.c. injection of Aβ1-42 for 8 months in Tg2576 and PSAPP mice did not reverse early-onset balance beam impairment nor effect spontaneous alternation in a Y-maze; vaccinated mice performed similarly to control tg mice.[9]

More recently, two papers have reported stunning behavioral improvement within very short time periods following passive transfer of Aβ-specific monoclonal antibodies (Mab). First, Dodart et al showed that weekly passive transfer of Mab 266 ($Aβ_{13-28}$) for 6 weeks led to improved object recognition and holeboard learning and memory 3 days after the last i.p. injection in 2 year old PDAPP tg mice.[10] At 24 months, PDAPP brains are loaded with Aβ deposition; however, in spite of the improved behavior, brain Aβ levels were unchanged. In another experiment, they found a dose-dependent improvement in the same tests in 11 month old PDAPP mice 24 hours after a single i.p. injection with Mab 266. Aβ levels in blood were increased and complexes of Aβ/anti-Aβ antibodies were detected in both blood and CSF. The authors concluded that because the behavioral deficits observed in PDAPP mice are reversed by binding of soluble Aβ, plaque deposition is unlikely to be solely responsible for behavioral changes in these mice. In another report, Kotilinek et al found that passive transfer of Mab BAM-10 (Aβ$_{1-12}$) in Tg2576 APP tg mice, given in 4 i.p. injections over 12 days, could reverse behavioral deficits in a spatial reference memory version of the Morris water maze without significantly reducing cerebral Aβ.[11] These authors suggested that the anti-Aβ antibodies bind small soluble Aβ assemblies responsible for the cognitive decline in APP tg mice. Both studies provide optimism for the potential of anti-Aβ therapy in improving cognition even in the context of advance plaque deposition in humans. However, such optimism is tempered by the fact that APP tg mice do not have the blatant neuron loss or neurofibrillary tangle formation observed in humans with AD. Therefore, one might predict somewhat less dramatic improvement in humans, depending on the level of structural neuronal damage at the time of treatment.

Characterization of the Mouse Immune Response to Aβ Vaccination

Combined results from many labs, including ours, have led to some general information on the immune response to immunization with Aβ peptides in WT and APP tg mice. For example, we showed that the genetic background of a mouse can influence its immune response to Aβ.[12] Following 12 weekly treatments with Aβ by i.p. injection, i.n. administration, or combination of the two, B6D2F1 mice generated antibodies earlier and in significantly higher quantities than did C57BL/6 mice, as shown in Figure 12.1. The combination treatment, as well as i.p. injection alone, produced the highest anti-Aβ titers and far surpassed those of mice treated with i.n. Aβ without adjuvant. These results demonstrate that background strain can regulate the immune response to Aβ vaccination and therefore have implications for Aβ vaccine experiments on APP and PSAPP tg mice of different genetic backgrounds.

Immune hyporesponsiveness in tg mice bearing mutant human APP and immunized once with human Aβ was reported by Monsonego et al.[13] However, such hyporesponsiveness was overcome by conjugating Aβ peptide with a carrier protein, bovine serum albumin (BSA). In a study by Wilcock et al, a similar hyporesponsiveness was observed in PSAPP mice compared to WT mice after three inoculations with Aβ; however, anti-Aβ titers were indistinguishable between PSAPP tg and WT mice after 9 inoculations.[14] Preliminary results from a long-term ongoing study in our lab indicates that the same is true for another APP tg mouse model, J20[15] mice. After immunizing both APP tg mice and their non-tg littermates with a single i.p. injection of Aβ followed by chronic i.n. boosting with Aβ + adjuvant LT(R192G) from 1 to 7 months of age, we found similar levels of anti-Aβ titers between tg and WT mice; however, the APP tg mice took longer to generate an anti-Aβ response.[16] Initially, the finding of immune hyporesponsiveness in APP tg mice caused concern that older humans, especially those with increased Aβ levels such as AD patients, may not be able to generate Aβ-specific antibodies with Aβ vaccination. However, the use of carrier proteins, adjuvants, and chronic immunizations may help alleviate this problem.

The humoral and cell-mediated immune characteristics of Aβ immunization in WT and APP/PSAPP mice have also been studied. The B cell epitope in mice has been shown to reside within the amino-terminus of Aβ. B cell epitopes have been reported within Aβ$_{1-12}$,[17] Aβ$_{1-15}$,[12,18-21] and Aβ$_{1-16}$.[6,22] In a presentation by Dr. Peter Seubert (Elan Corporation) at the International Alzheimer's Meeting in Stockholm in July, 2002, a B cell epitope of Aβ$_{2-7}$ was described for Aβ immunized mice. Like others, we have found that serum from Aβ immunized mice can be used immunohistochemically to label human AD plaques. Incubating the serum with Aβ$_{1-15}$ peptide, but not other overlapping peptides of Aβ, resulted in ablation of AD plaque IR as shown in Figure 12.2, a previously unpublished illustration from our original

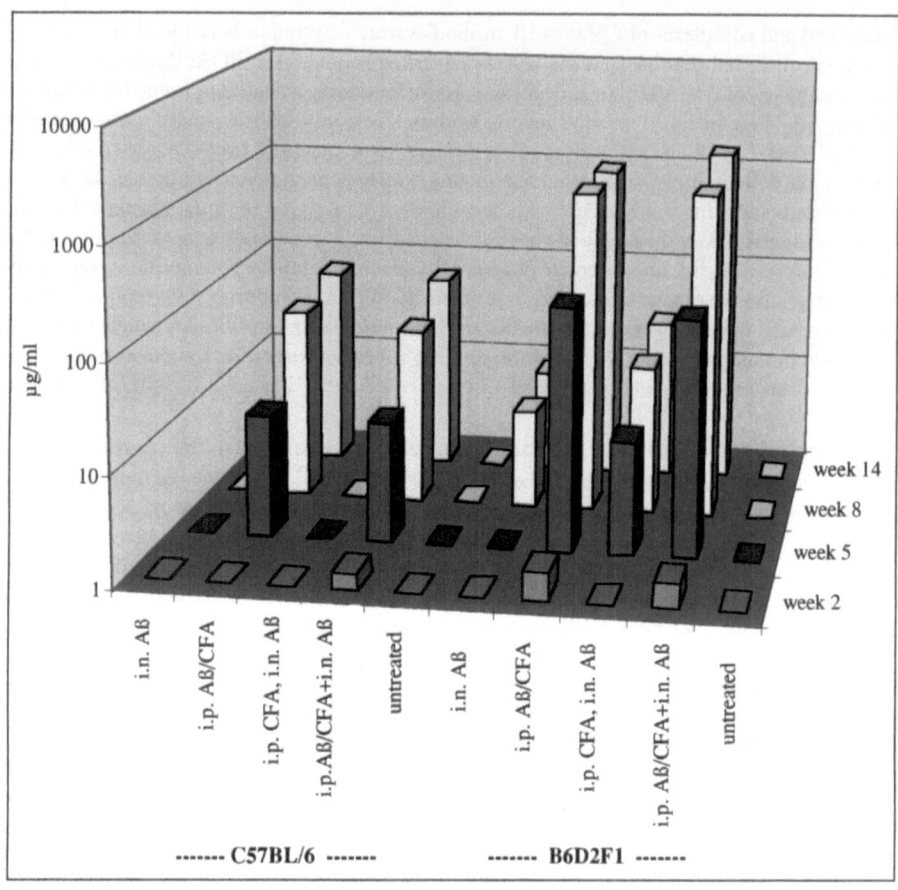

Figure 12.1. Aβ antibody titers were compared between two WT mouse strains (C57BL/6 and B6D2F1) using 4 different immunization protocols.
A cocktail of three parts Aβ1-40 and 1 part Aβ1-42 (total 100 μg) was used for all treatments. Mice were treated with i.n Aβ without adjuvant, i.p. Aβ + CFA, a single i.p. injection of CFA followed by chronic i.n. Aβ, or overlapping i.p Aβ/CFA plus i.n. Aβ, or left untreated for 12 weeks. Titers are reported as μg/ml; note Y-axis logarithmic scale. Further details and Aβ titer means are provided in reference 12. For each immunization protocol, B6D2F1 mice generated more Aβ antibodies. Aβ titers were highest in mice treated with the combined i.p. Aβ/CFA plus i.n. Aβ, followed closely by i.p. Aβ/CFA alone. Lower Aβ titers were detected in mice treated with i.p. CFA/i.n. nasal Aβ and i.n. Aβ alone. Thus, strain differences effect the ability of a mouse to generate an immune response to Aβ immunization.

intranasal Aβ study in PDAPP mice. Immunoglobulin (Ig) isotypes have been examined following Aβ immunization. In most studies, the predominant Ig isotype generated by Aβ immunization is IgG2b, followed by IgG1 and IgG2a, with lesser amounts of IgA and IgM.[3,6,17,18,22] Figure 12. 3 illustrates Ig isotype-specific human plaque labeling using sera from our first study[3,4] in which PDAPP mice were treated with intranasal Aβ. We have shown that in addition to enhancing antibody titers, adjuvants can also shift the relative amounts of Ig isotypes.[19]

In addition to B cell epitope mapping of anti-Aβ antibodies, Aβ-specific T cell epitopes have also been recently reported. Monsonego et al found strain differences in T cell epitopes in C57BL/6 (Aβ15-30) compared to SJL/J (Aβ9-24) mice following Aβ immunization.[23] In Dr.

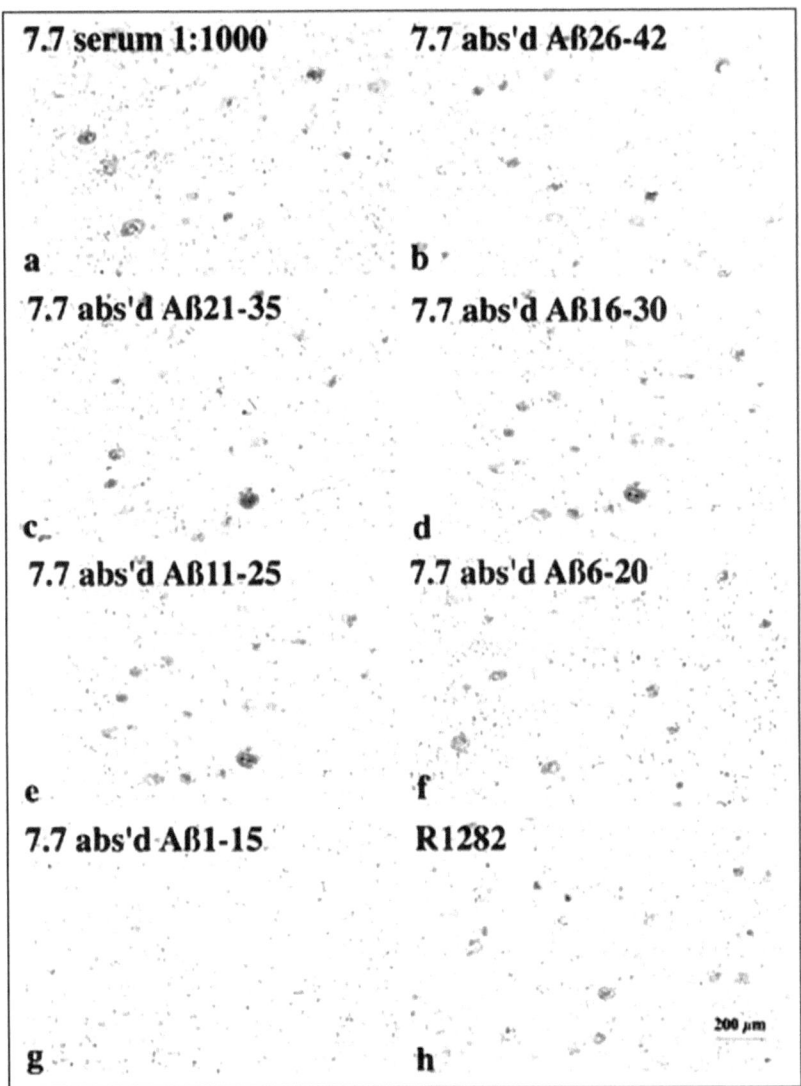

Figure 12.2. Aβ antibody epitope-mapping.
Serum from PDAPP mice immunized weekly by i.n. human Aβ1-40 from 5 to 12 months of age was used to immunolabel Aβ plaques in serial human AD brain sections. The mouse serum was pre-incubated with each of six overlapping 15-mer Aβ peptides. Only Aβ1-15 abolished plaque staining, indicating that the Aβ antibodies in the mouse serum recognized an epitope within Aβ1-15. Aβ Pab, R1282, was used as a positive control for plaque labeling. Further details of this study can be found in references 3 and 4. Scale bar: 200 microns.

Seubert's presentation at the Stockholm meeting, he described a T cell epitope of $Aβ_{14-18}$ in Aβ immunized mice. Aβ immunization in mice has led to a proliferative T cell response[22,23] and a shift in T cell response of splenocytes from WT and Tg2576 APP tg mice toward enhancing Th2 and decreasing Th1 immunity.[24]

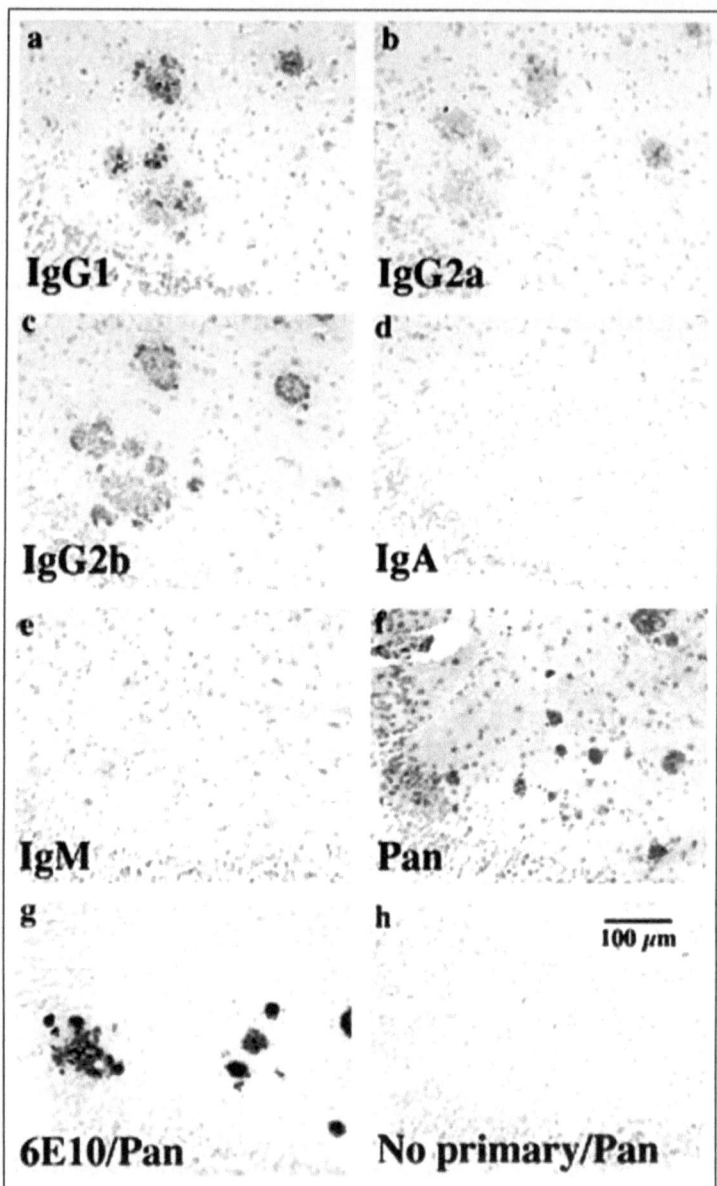

Figure 12.3. Anti-Aβ immunoglobulin isotyping.
Serum from PDAPP mice immunized weekly by i.n. human Aβ1-40 from 5 to 12 months of age was used as primary antibody to immunostain plaques in serial human AD brain sections. Biotinylated Ig isotype-specific secondary antibodies were used to detect the antibody isotypes generated by Aβ immunization. Strong plaque labeling was detected with IgG2b (c) and IgG1 (a) secondary antibodies, while weaker staining was observed using an IgG2a (b) antibody. Plaques were not reactive with IgM or IgA secondary antibodies. A pan-specific (IgG, IgM, IgA) secondary antibody and labeling with an Aβ Mab, 6E10, were used as positive controls. Omission of primary antibody produced no plaque labeling. An Ig isotype ELISA confirmed these results and revealed the most abundant isotype to be IgG2b, followed by IgG1 and then Ig2a. Further details of this study can be found in references 3 and 4. Scale bar: 100 microns.

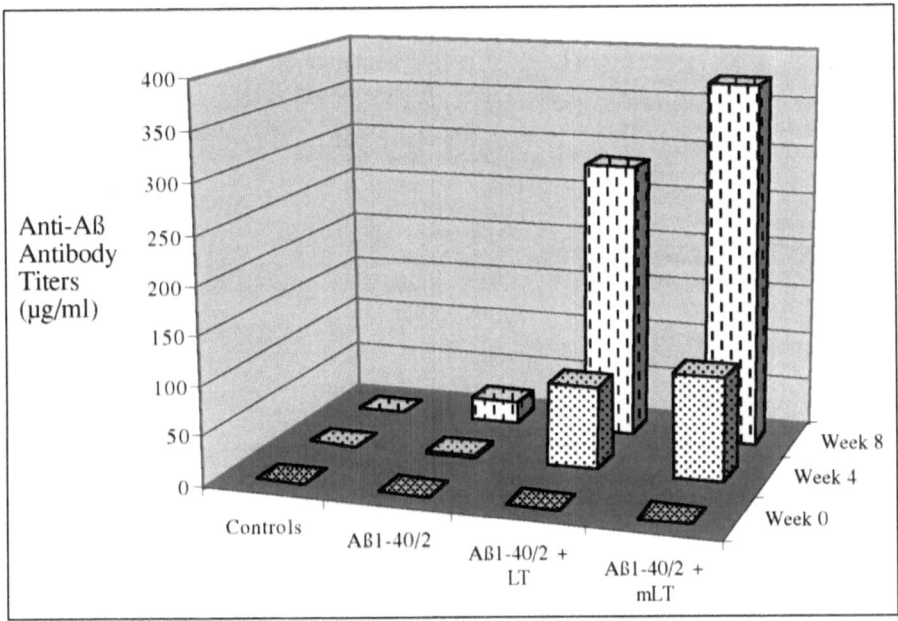

Figure 12.4. Mucosal adjuvants, LT and LT(R192G) enhance anti-Aβ titers.
Anti-Aβ antibody titers were determined by ELISA for the sera of B6D2F1 mice immunized twice weekly with i.n. Aβ without adjuvant, i.n. Aβ with the adjuvant heat labile enterotoxin, LT, i.n. Aβ with the nontoxic, mutant form of the adjuvant, LT(R912G), or left untreated for 8 weeks. Anti-Aβ antibodies were not detected in any of the pre-immune sera but their numbers increased from week 4 to week 8 in all groups given i.n. Aβ. By week 8, a 12-fold increase in Aβ titers was found serum from mice treated with i.n. Aβ + LT compared to those in mice treated without adjuvant. A 16-fold increase in Aβ titers was detected in mice treated with i.n. Aβ + LT(T192G) compared to those in mice treated without adjuvant. Further details of this study can be found in reference 19.

Immunization Strategies in Mice

The most common method for Aβ immunization in mice has been chronic i.p. or s.c. injection of full-length human Aβ peptide using Complete Freund adjuvant (CFA) for the first injection, Incomplete Freund adjuvant (IFA) for the next few injections and the peptide alone, hereafter.[1,5-8,17] Intranasal (i.n.) administration has been shown to be effective in generating anti-Aβ titers by us[3,4] and may be more convenient than injections for patients as it can be self-administered and does not require sterile instruments nor travel for office visits to receive treatment. Anti-Aβ titers were dramatically increased using a mucosal adjuvant, *E. coli* heat labile enterotoxin, LT, or a mutant form, LT(R192G), which has significantly reduced toxicity.[19] Figure 12.4 illustrates a 12-fold increase in Aβ titers in WT mice using LT and a 16-fold increase in Aβ titers using LT(R192G) with i.n. Aβ immunization after 8 weeks compared to untreated WT mice. The adjuvants LT and LT(R192G) produced no obvious toxic effects in the mice and were well-tolerated.

We have since developed a prime/boost protocol in which a single i.p. injection of Aβ + CFA is followed by weekly i.n. boosting with Aβ + LT or LT(R192G). This immunization strategy has proven very effective in raising anti-Aβ titers, lowering cerebral Aβ and increasing peripheral Aβ after only 8 weeks in young PSAPP mice.[21] Figure 12.5 depicts the significant reduction observed in the number of plaques formed in both frontal cortex and hippocampus

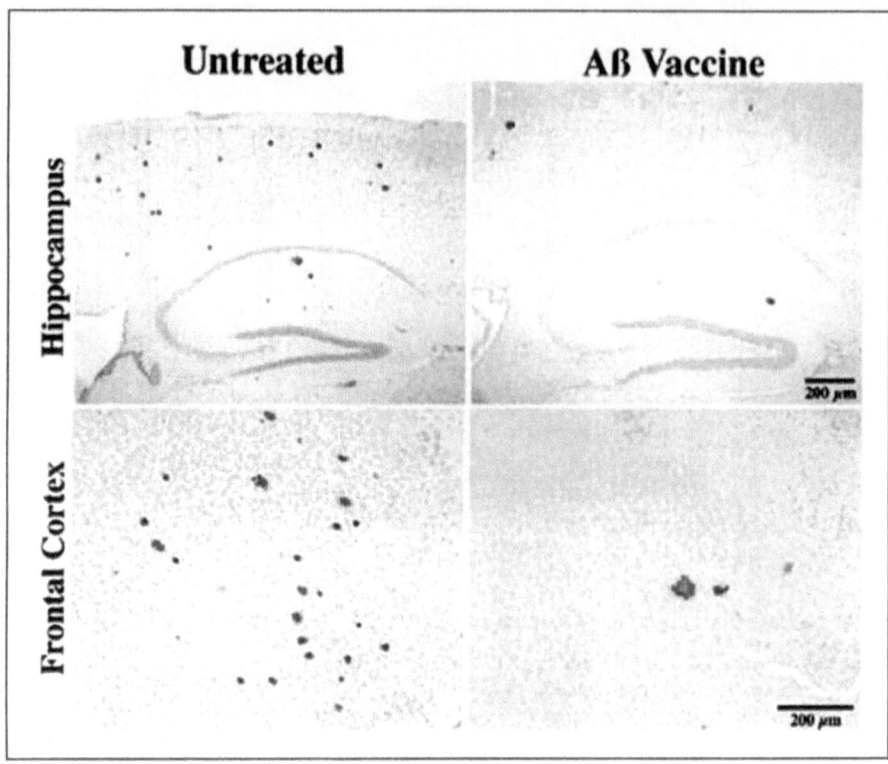

Figure 12.5. After only 8 weeks of Aβ immunization in PSAPP tg mice, the number of Aβ plaques was reduced by 75%.

Mice were given one i.p. injection of Aβ40/42 + CFA and then chronically boosted with i.n. Aβ + LT (2x/week) for 8 weeks. Treatments began at 5 weeks of age, prior to plaque deposition in this accelerated model Aβ pathogenesis, and lasted 8 weeks. ELISA of guanidine-HCL-extracted brain homogenates revealed a significant 58% decrease in Aβ1-42. For more details, see reference 21. Scale bars: 200 microns.

in these mice. In addition, we have found that this prime/boost regimen is also effective in generating anti-Aβ titers in WT mice using full-length Aβ to prime by i.p. injection and boosting weekly with a shorter peptide, Aβ1-15, with LT.[20] The anti-Aβ antibody titers (~200 μg/ml) were not as high as those using full-length Aβ peptide throughout vaccination but were roughly the same levels (~240 μg/ml) shown to lower cerebral Aβ in our PSAPP study. Aβ1-15 peptide has the benefit of being less costly to produce and may have fewer toxic effects as it may not be recognized as an endogenous Aβ species or a T cell epitope.

Various other modifications of Aβ peptide for immunization have been tested with the goal of improving the immune response and/or reducing the potential toxic effects of immunizing with full-length, aggregated Aβ peptides. For example, Sigurdsson et al reported that monthly injections of a soluble, nonamyloidogenic, nontoxic form of Aβ, Aβ$_{1-30}$, modified by the addition of six lysine residues at the N-terminus to keep the peptide soluble and prevent fibril formation, led to a significant reduction in brain Aβ levels in Tg2576 APP mice.[25] This was the first study to demonstrate the effective use of nonaggregated Aβ as an immunogen to generate anti-Aβ titers and reduce brain Aβ levels in APP tg mice. In a study by Lambert et al, small soluble Aβ oligomers (known as protofibrils or ADDLs) were used successfully as immunogens in rabbits to generate Aβ-specific antibodies that recognized assemblies of Aβ biochemically and prevented the toxicity of soluble oligomers in cell culture.[26] Because these small Aβ oligomers are known to have CNS toxicity,[27] directing a vaccine against them or administering anti-Aβ antibodies specific for them via passive transfer provides an attractive approach to lowering the levels of such Aβ species.[28] Another innovative approach was taken by Frenkel et al in which human Aβ$_{3-6}$ (EFRH), a region important for Aβ fibril formation, was displayed on a filamentous phage and used to immunize WT mice and guinea pigs.[29] Anti-aggregating Aβ antibodies were generated in both animal species. Importantly, because guinea pigs share the same Aβ amino acid sequence as humans, the antibodies generated by the EFRH phage vaccine were autoantibodies yet did not induce any obvious autoimmune toxicity.[30]

Passive transfer of certain, but not all, Aβ antibodies by peripheral injection in APP tg mice has been shown to reduce cerebral Aβ levels and plaque burden,[2] as well as increase Aβ levels in the blood,[31] and, to improve behavioral testing results, even after very brief periods of treatment in which cerebral Aβ levels remained unchanged, as described earlier in this chapter.[10,11] Recently, Mohajeri et al reported that passive transfer of Aβ antibodies to APP tg mice protected them from neuronal loss following seizure induction and reduced brain Aβ levels.[32] In their study, APP tg mouse neurons showed an elevated vulnerability to seizure activity compared to those of WT mice. Treating the APP tg mice with passive transfer of Aβ antibodies at an age prior to plaque formation, but while Aβ levels were rising, protected the neurons.

Direct passive transfer of Aβ antibodies into brain has been demonstrated via intracerebral ventricular injection (i.c.v.) in APP tg mice and a mouse model of aging, SAMP8. Chuahan et al administered a single i.c.v. injection of an anti-Aβ antibody (Mab AMY-33, against Aβ$_{1-28}$) into the third ventricle of 10 month old Tg2576 APP mice, prior to the onset of robust plaque formation, and one month later found that they had significantly restored the levels of SNAP-25, a presynaptic nerve terminal protein involved in synaptic vesicle exocytosis and neurotransmitter release, which typically drops in these mice between 10 and 11 months of age.[33] In addition, rising GFAP levels, a marker of astrogliosis, in the treated APP tg mice were restored to WT levels. In another report, Morley et al administered Aβ monoclonal or polyclonal antibodies via i.c.v. injection to the third ventricle in 12 month old SAMP8 mice, a mouse model of accelerated senescence with Aβ overproduction, and showed improved acquisition and memory in an aversive T maze test when injected 1-14 days prior to testing.[34] Their results strongly suggest a relationship between elevated Aβ levels and loss of acquisition and retention in SAMP8 mice.

Proposed Mechanisms for the Beneficial Effects of Aβ Vaccination in Mice

Thus far, three major mechanisms have been proposed for the Aβ clearing and behavioral improvement effects observed following either active or passive Aβ immunization. It is possible that all three may play roles in modulating Aβ following vaccination. The first mechanism proposed was based upon reports by Solomon et al in 1996 and 1997 showing that monoclonal antibodies directed against the amino-terminus of Aβ prevented fibrillar aggregation of Aβ in vitro,[35] disaggregated preformed Aβ fibrils, and prevented their toxicity in PC12 cells.[36] A confirmation of this finding in vivo was reported by Bacskai et al in 2001, in studies in which live imaging of Aβ plaques using multiphoton microscopy was employed before and after topical application of Aβ mAb, 10D5 (Aβ3-6), to the surface of the cortex in 20 month old PDAPP tg mice.[37] After 3 days, 70% of plaques (45 of 65) within the viewing range had been cleared while the remaining plaques were frequently associated with activated microglia.

These results were also consistent with another proposed mechanism, Fc-receptor mediated phagocytosis of Aβ, first described in ex vivo studies on brain sections from PDAPP tg mice and humans with AD by Bard et al in which Aβ antibodies applied directly to the sections induced microglia to phagocytose plaque Aβ.[2] Data in support of this mechanism include a report by Wilcock et al in which a transient rise in CD45-IR microglial activation was observed after five inoculations in PSAPP tg mice but was equivalent to that in control injected tg mice after nine inoculations.[14] In our own studies, we have consistently found decreased numbers of activated microglial cells in Aβ immunized APP or PSAPP tg mice with significantly reduced plaque burdens; however, we have only examined the brains at the end of each study.[3,4,21] It is possible that a transient rise in microglial activation occurred during the early stages of immunization treatments and then gradually was diminished by the end of each study. Using confocal microscopy and double-labeling with anti-CD45 (for microglia) and R1282 (a general Aβ pAb, gift of Dr. Dennis Selkoe), we have been unable to distinguish plaque-associated microglia from immunized vs. nonimmunized PSAPP mice, as shown in Figure 12.6. In addition, it is implicit in this proposed mechanism that the Aβ antibodies cross the blood-brain-barrier (BBB). Bard et al[2] have reported the presence of passively transferred Aβ antibodies, given by i.p. injection, in the brains of PDAPP Tg mice; however, other researchers, such as DeMattos et al[31] and our group[21] have not been able to visualize bound Aβ antibodies in brain parenchyma after either passive transfer or active Aβ immunization. Differences in tissue fixation and staining protocols may explain some of these discrepancies. Lastly, in a very recent report by Bacskai et al, topical cortical application of the FITC-labeled whole mAb, 3D6 (Aβ$_{1-5}$) or just the FITC-labeled F(ab')2 fragments of 3D6 missing the Fc region of the antibody, led to roughly equivalent clearance (50% and 45%, respectively) of plaques from 18 month old Tg2576 APP tg mice and 20 month old PDAPP tg mice within 3 days as determined by multiphoton imaging.[38] This result argues that Fc-receptor binding by Aβ antibodies is not solely responsible for the clearance of Aβ and suggests that alternative or additional mechanisms, such as dissaggregation of Aβ, are likely to play a role.

A third mechanism proposed to explain the beneficial effects of the Aβ vaccine in APP and PSAPP mice was first described by DeMattos et al in 2001 in a report showing that peripheral passive transfer of an Aβ Mab, 266 (Aβ$_{13-28}$), altered CNS and plasma Aβ clearance while lowering cerebral Aβ in PDAPP Tg mice.[31] A 1000-fold increase in plasma Aβ was observed several days after intravenous (i.v.) injection of Mab 266, suggesting that the Aβ antibody was acting as a "sink" in the periphery by enhancing efflux of Aβ from the brain to the peripheral compartments for clearance. Further discussion of this mechanism and the balance of Aβ equilibrium between CNS and blood are provided elsewhere in this book, in a chapter written by Drs. DeMattos and Holtzman. Our data from Aβ immunization in PSAPP tg mice support the findings of DeMattos and colleagues. After 8 weeks of chronic, active Aβ immunization, we found a significant reduction in cerebral Aβ and a concomitant significant increase in serum Aβ.[21] As in the 2001 report by DeMattos et al,[31] in our study most of the peripheral Aβ bound to Aβ antibodies, forming immune complexes. This was the first evidence that active immunization increased serum Aβ, suggesting that the antibodies enhanced Aβ clearance from the brain. As mentioned earlier, behavioral studies by Dodart et al reported a rapid improvement in two tests of spatial memory following passive transfer with Mab 266 in PDAPP mice; a rapid rise in plasma Aβ was found to correlate with these behavioral changes.[10] Thus, while the different mechanisms that have been proposed to lower cerebral Aβ following Aβ immunization may work together, it is likely that the efflux of soluble Aβ from brain to periphery, leading to Aβ immune complexes in the blood, plays an important role. Interestingly, a recent report by Pan et al demonstrated that Aβ antibody/Aβ immune complexes decreased the passage of Aβ from blood to brain.[39] Incubation of ^{125}I-Aβ$_{1-40}$ with either of two Aβ Mabs (3D6 or mc1) prior to i.v. injection led to reduced influx of Aβ$_{1-40}$ into the brain in WT mice. These results suggest that Aβ antibodies may also play a role in preventing Aβ from crossing the BBB and depositing in brain.

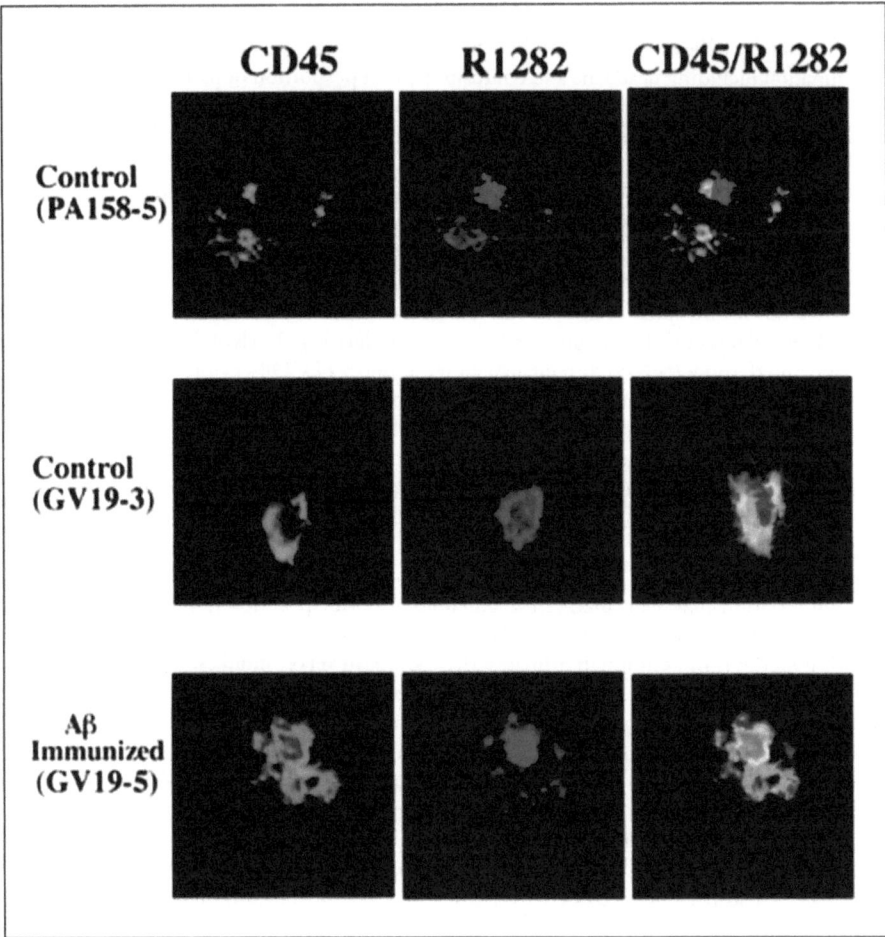

Figure 12.6. Double immunofluorescent labeling of plaques (R1282) and microglia (CD45) showed similar patterns of colocalization in Aβ immunized (bottom) and untreated (top and middle) PSAPP mice after 8 weeks of Aβ immunization, as described in reference 21.

As illustrated in Figure 12.5, a dramatic reduction in plaque burden was seen in the 13 week old mice after treatment. CD45-immunoreactive microglia colocalized with compacted plaques in treated and untreated mice; however, because the numbers of plaques were significantly reduced in the Aβ immunized mice, the number of labeled microglia was also reduced. Images were obtained using a Zeiss Axiovert 100 M laser-scanning confocal microscope (LSM510).

Aβ Vaccination in Human Clinical Trials

Elan Corporation and Wyeth/Ayerst/American Home Products successfully completed their Phase 1 safety studies in July, 2001 in which more than 100 patients with mild-to-moderate AD in the US and UK were immunized with either a single or escalating doses of AN-1792, a synthetic human Aβ$_{1-42}$ peptide. Previously, Schenk et al had shown AN-1792 to be effective in clearing cerebral Aβ following immunization in PDAPP tg mice.[1] Although not all patients developed anti-Aβ titers, no adverse effects were observed and the clinical program moved into the next stage. A Phase IIa clinical trial was initiated in the Fall of 2001, in which 375 patients with mild-moderate AD were recruited at 13 centers in Europe and the USA. A total of 300 patients received one to several injections of AN-1792 along with an adjuvant, QS-21 (saponin). By March 1, 2002, 15 patients who had received the Aβ vaccine developed signs of CNS inflammation including headache and confusion; only a proportion of patients had developed anti-Aβ titers at the time (www.Elan.com/NewsRoom/NewsYear2002). Due to these adverse events, the dosing of all patients was stopped. Recently, Dr. Schenk reported that number of patients with menigoencephilitis has risen to 17 of 300. [40] The exact cause for these untoward effects remains unknown. Speculation regarding the explanation of these adverse events includes an autoimmune-like T cell activation stimulated by the human Aβ peptide, toxicity due to the adjuvant, or viral infection in the CSF. In 2000, Grubeck-Loebenstein et al warned that immunization of humans with synthetic Aβ peptide may lead to activation of Aβ-specific T cells and a downward spiral of unwanted inflammatory events targeted at cells overproducing Aβ.[41] Marx et al reported that Aβ-specific T cell lines produce high levels of the pro-inflammatory cytokine, interferon γ (IFN-γ) and include a high number of CD8+ cytotoxic T lymphocytes (CTL) that have the potential to lyse cells in the brain if permitted entry through the BBB.[42,43] Munch et al suggested that perhaps Aβ-specific T cells might also target APP (which contains the Aβ antigenic determinant) and/or Aβ antibodies might bind APP on the cell surface of neurons, trigger complement activation and opsonization, both potentially leading to damage in the brain.[44] However, Hock et al recently reported that the Aβ-specific antibodies obtained from human patients in the Swiss clinical test site did not recognize either full-length APP or C-terminal APP fragments.[45] Examining T cell reactivity to Aβ peptide both before and after Aβ immunization may help predict the outcome of both Aβ antibody production and the potential for a CTL response and pro-inflammatory cytokine induction. We are currently testing this hypothesis in nonhuman primates, the Caribbean vervet monkey.

Another possible cause for the adverse effects observed in ~4% of AD patients in the Elan trial derives from the use of the adjuvant, QS-21, a highly purified saponin obtained from the bark of the *Quillaja saponaria* Molina tree, in conjunction with AN-1792.[46] Adjuvant QS-21 has been shown to enhance antibody generation, augment both Th1 and Th2 cytokine production and induce a potent CTL response to the antigen with which it is administered.[47,48] However, as suggested by Singh et al, one must consider the level of potency as an adjuvant to enhance antibody titers versus the ability of a CTL response to induce adverse effects.[49] As of 2000, QS-21 had been given by either intramuscular or subcutaneous injection in over 2200 human subjects participating in more than 50 clinical trials; the only reported side effect was transient pain at the site of injection.[50] Although some additional short-term side effects such as fever and chills[51,52] as well as vasovagal episodes and hypertension[53] have been reported, in the majority of studies, QS-21 has been deemed relatively safe for use in humans, including the elderly (e.g., refs. 52 and 54).

Lastly, viral load was found in CSF in at least one of the four earliest patients with adverse effects. However, it is unlikely that viral infection will be found in all 17 patients who developed neurological symptoms. Some have speculated on the slight chance of contamination at the site of lumbar puncture leading to either bacterial or viral infection, although, it is unclear when the punctures were done and whether they might have relevance to the neurologic symptoms observed. Again, until the clinical data regarding the patients with adverse effects in the Elan trial are revealed, one can only speculate as to the exact cause of neurological inflammation in those patients.

Future Strategies in AD Vaccine Development

With the halting of the first human clinical trial for Aβ immunization last March, the emphasis for the future of AD immunotherapy relies upon creating a safe means by which to sequester Aβ proteins without inducing a toxic effect within the CNS. A list of potential therapeutic strategies is already beginning to form. First, passive transfer of Aβ-specific antibodies, as shown to be effective in lowering cerebral Aβ[2] and improving behavior[10] in PDAPP mice, is likely to avoid the generation of an Aβ-specific T cell response and, in addition, does not require the use of an adjuvant. Aβ antibodies may be delivered by i.m. or s.c. injection, or by i.c.v. delivery, though the latter would likely be more difficult to employ in humans. Poduslo et al have shown that modifying an Aβ$_{42}$-specific radio-labeled Mab with polyamine increased its permeability from blood to brain 36-fold, suggesting that polyamine modifications may enhance the passage of Aβ antibodies to brain via peripheral passive immunization in humans.[55] Depending upon the mechanism for the adverse events in the Elan trial, this may be beneficial in that enhancing antibody entry into the brain may solubilize and sequester more Aβ, or it may be deleterious if the antibodies fix complement and induce the complement cascade to attack neurons. Aβ antibodies specifically recognizing toxic, small soluble oligomers of Aβ have been raised in rabbits.[26] Passive immunization with humanized versions of such antibodies may lower the levels of toxic, soluble assemblies of Aβ[28] and have a marked effect on cognition and memory as suggested by studies in mice.[10,11] In a recent report by Dodel et al, intravenous immunoglobulin (IVIG) administration of purified Aβ antibodies (from commercially available human IVIGs) over 3 consecutive days in seven non-AD patients with neurological or non-neurological immune-mediated conditions (mean age, 62.7 years) in Germany, led to increased concentrations of Aβ antibodies in CSF and serum, decreased levels of Aβ (total and 1-42) in CSF, and increased total Aβ levels in serum one-to-three weeks later.[56] Their results support and confirm the peripheral "sink" hypothesis[31] by showing that infusion of Aβ antibodies outside the BBB may enhance clearance of soluble Aβ peptides from the CNS and suggest that infusion with IVIG-purified Aβ antibodies may be a potential therapeutic strategy for AD.

Active immunization strategies using novel, nontoxic formulations for Aβ antigen presentation are also being pursued. Brayeden et al reported successful antibody generation using Aβ1-42 delivered inside of poly(lactide-co-glycolide) (PLG) microspheres via i.p. or s.c. injection in WT mice.[57] The benefits of using Aβ-loaded PLG microspheres are that no other adjuvant is required, a predominantly Th1 response is observed (which may enhance antibody titers, but its cell-mediated effects remain unclear) and, under certain formulations, the number of immunizations necessary to maintain an efficacious titer level may be reduced. Nicolau et al reported that i.p. injection of palmitoylated Aβ$_{1-16}$ peptide reconstituted in liposomes containing lipid A and alum generated a strong immune response in WT mice and NORBA tg mice, which overexpress human APP leading to Aβ plaque deposition in pancreas and, both prevented plaque formation in young NORBA mice and significantly reduced plaques in older NORBA mice.[58] Vaccination using a filamentous phage vector that displays a small epitope of Aβ known to be important for Aβ fibril formation and aggregation, human Aβ$_{3-6}$ (or EFRH), has been tested by Frenkel et al. This strategy was found to be effective in generating antibodies in both mice and guinea pigs; guinea pigs share the same Aβ sequence as humans, therefore, the result of such immunization is the production of Aβ autoantibodies.[29,30]

Several other vaccine strategies were reported recently at the International AD Meeting in Stockholm in July, 2002. Cao et al found that using recombinant Aβ proteins expressed in *E. coli* led to higher anti-Aβ titers and at earlier time points when compared to using synthetic Aβ peptide as the immunogen.[59] Naked DNA vaccines targeted at enhancing anti-Aβ titers were reported using Aβ fused with IL-4 or 3 copies of C3d,[60] or Aβ fused with BRI, a protein involved in amyloid deposition in Familial British dementia.[61] The fusion of immunopotentiators with Aβ in DNA vaccines may help overcome the lack of immune response observed in some patients in the Elan Phase I and IIa trials.

Directing the T cell response to enhance CD4+ T cells for help in B cell generation of Aβ-specific antibodies and to suppress CD8+ Aβ-directed CTLs may be desirable in a human AD vaccine.[42] The use of adjuvants, either alone or in combination, can direct such an immune response. For example, while QS-21 generates IgG2a/2b isotypes (Th1/Th2) in immunization with antigen + conjugate, QS21-alum mixture leads to a more Th2-directed response, with predominant IgG1/IgG2b isotype antibodies being generated.[62] The use of conjugates in vaccines, such as keyhole limpet hemacyanin (KLH), may be useful in the AD vaccine. In studies of human cancer antigens in mice, the predominant T-cell immune response was induced against KLH, leading to higher levels of antibody against the conjugated antigens.[63] Certain cytokines, such as IL-1α, IL-12, IL-18, and IFNγ have been used successfully as adjuvants to enhance antibody titers against a co-administered antigen.[64,65] Shifts in Th1 to Th2 responses have been demonstrated using IL-12 as an adjuvant for an oral combined vaccine of tetanus toxin and cholera toxin compare to those induced by intranasal use of IL-12 as an adjuvant.[66]

While the adverse neurologic effects observed in 17 of 360 patients immunized with AN-1792 ($Aβ_{1-42}$) and QS-21 in the Elan trial are being investigated and taken quite seriously, the search for a safe, effective vaccine still exists. We, like others, believe that the development of a nontoxic, efficacious form of Aβ immunotherapy is a realistic possibility. Whether this will involve passive immunization of Aβ-specific antibodies or active immunization using modified/encapsulated Aβ, adjuvants, and/or DNA vaccines remains to be determined. However, we are optimistic that there is still today a great potential for prevention and/or treatment of Alzheimer's disease via Aβ immunotherapy in the future.

References

1. Schenk D, Barbour R, Dunn W et al. Immunization with amyloid-β attenuates Alzheimer-disease-like pathology in the PDAPP mouse. Nature 1999; 400:173-177.
2. Bard F, Cannon C, Barbour R et al. Peripherally administered antibodies against amyloid beta-peptide enter the central nervous system and reduce pathology in a mouse model of Alzheimer disease. Nat Med 2000; 6(8):916-919.
3. Weiner HL, Lemere CA, Maron R et al. Nasal administration of amyloid-beta peptide decreases cerebral amyloid burden in a mouse model of Alzheimer's disease. Ann Neurol 2000; 48(4):567-579.
4. Lemere CA, Maron R, Spooner ET et al. Nasal Aβ treatment induces anti-Aβ antibody production and decreases cerebral amyloid burden in PD-APP mice. Ann NY Acad Sci 2000; 920:328-331.
5. Vehmas AK, Borchelt DR, Price DL et al. β-amyloid peptide vaccination results in marked changes in serum and brain Aβ levels in APPswe/PS1ΔE9 mice, as detected by SELDI-TOF-based ProteinChip technology. DNA and Cell Biology 2001; 20(11):713-721.
6. Das P, Murphy MP, Younkin LH et al. Reduced effectiveness of Aβ1-42 immunization in APP transgenic mice with significant amyloid deposition. Neurobiol Aging 2001; 22:721-727.
7. Janus C, Pearson J, McLaurin J et al. A beta peptide immunization reduces behavioural impairment and plaques in a model of Alzheimer's disease. Nature 2000; 408(6815):979-982.
8. Morgan D, Diamond DM, Gottschall PE et al. A beta peptide vaccination prevents memory loss in an animal model of Alzheimer's disease. Nature 2000; 408(6815):982-985.
9. Arendash GW, Gordon MN, Diamond DM et al. Behavioral assessment of Alzheimer's transgenic mice following long-term Aβ vaccination: task specificity and correlations between Aβ deposition and spatial memory. DNA and Cell Biology 2001; 20(11):737-744.
10. Dodart J-C, Bales KR, Gannon KS et al. Immunization reverses memory deficits without reducing brain Aβ burden in Alzheimer's disease model. Nature Neurosci 2002; 5(5):452-457.
11. Kotilinek LA, Bacskai B, Westerman M et al. Reversible memory loss in a transgenic model of Alzheimer's disease. J Neurosci 2002; 22(15):6331-6335.
12. Spooner ET, Desai RV, Mori C et al. The generation and characterization of potentially therapeutic Aβ antibodies in mice: differences according to strain and immunization protocol. Vaccine 2002; 21:290-297.
13. Monsonego A, Maron R, Zota V et al. Immune hyporesponsiveness to amyloid-β peptide in APP transgenic mice: Implications for the pathogenesis and treatment of Alzheimer's disease. Proc Natl Acad Sci USA 2001; 98:10273-10278.

14. Wilcock DM, Gordon, MN, Ugen KE et al. Number of Aβ inoculations in APP+PS1 transgenic mice influences antibody titers, microglial activation and congophilic plaque levels. DNA and Cell Biology 2001; 20(11):731-736.

15. Mucke L, Masliah E, Yu G-Q et al. High-level neuronal expression of Aβ1-42 in wild-type human amyloid protein precursor transgenic mice: synaptotoxicity without plaque formation. J Neurosci 2000; 20(11):4050-4058.

16. Lemere C, Spooner E, LaFrancois J et al. Evidence for the "peripheral sink" hypothesis following chronic, active Aβ immunization in PSAPP mice. Neurobiol Aging 2002; Abstr. 401:S106.

17. Town T, Tan J, Sansone N et al. Characterization of murine immunoglobulin G antibodies against human amyloid-β1-42. Neuroscience Letters 2001; 307:101-104.

18. Lemere CA, Maron R, Selkoe DJ et al. Nasal vaccination with β-amyloid peptide for the treatment of Alzheimer's disease. DNA and Cell Biology 2001; 20(11):705-711.

19. Lemere CA, Spooner ET, Leverone JF et al. Intranasal immunotherapy for the treatment of Alzheimer's disease: *Escherichia coli* LT and LT(R192G) as mucosal adjuvants. Neurobiology of Aging 2002: 23(6):991-1000.

20. Leverone JF, Spooner ET, Lehmann H et al. Aβ1-15 is less immunogenic than Aβ1-40/42 for intranasal immunization of wild-type mice but may be effective for boosting. Vaccine 2002; in press.

21. Lemere CA, Spooner ET, LaFrancois J et al. Evidence for peripheral clearance of Aβ following chronic, active Aβ immunization in PSAPP mice. (submitted).

22. Dickey CA, Morgan DG, Kudchodkar S et al. Duration and specificity of humoral immune responses in mice vaccinated with the Alzheimer's disease-associated β-amyloid 1-42 peptide. DNA and Cell Biology 2001; 20(11):723-729.

23. Monsonego A, Zota V, Selkoe D et al. Immunogenic aspects of amyloid-β peptide: Implications for pathogenesis and treatment of Alzheimer's disease. Neurobiol Aging 2002; Abstr. 503:S133.

24. Town T, Vendrame M, Patel A et al. Reduced Th1 and enhanced Th2 immunity after immunization with Alzheimer's β-amyloid1-42. J Neuroimmunol 2002; 132:49-59.

25. Sigurdsson EM, Scholtzova H, Mehta PD et al. Immunization with a nontoxic/nonfibrillar amyloid-β homologous peptide reduces Alzheimer's disease-associated pathology in transgenic mice. Am J Pathol 2001; 159:439-447.

26. Lambert MB , Viola KL, Chromy BA et al. Vaccination with soluble Aβ oligomers generates toxicity-neutralizing antibodies. J Neurochem 2001; 79:595-605.

27. Lambert MP, Barlow AK, Chromy BA et al. Diffusible, nonfibrillar ligands derived from Aβ$_{1-42}$ are potent central nervous system neurotoxins. Proc Natl Acad Sci USA 1998; 95:6448-6453.

28. Klein W, Abeta toxicity in Alzheimer's disease: globular oligomers (ADDLs) as new vaccine and drug targets. Neurochem Int 2002; 41(5):345-352.

29. Frenkel D, Katz O, Solomon B. Immunization against Alzheimer's β-amyloid plaques via EFRH phage administration. Proc Natl Acad Sci USA 2000; 97(21):11455-11459.

30. Frenkel D, Kariv N, Solomon, B. Generation of auto-antibodies towards Alzheimer's disease vaccination. Vaccine 2001; 19:2615-2619.

31. DeMattos RB, Bale KR, Cummins DJ et al. Peripheral anti-Aβ antibody alters CNS and plasma clearance and decreases brain Aβ burden in a mouse model of Alzheimer's disease. Proc Natl Acad Sci USA 2001; 98:8850-8855.

32. Mohajeri MH, Saini K, Schultz JG et al. Passive immunization against β-amyloid peptide protects CNS neurons from increased vulnerability associated with an Alzheimer's disease-causing mutation. J Biol Chem Papers in Press: M203193200 2002; June 14, 2002.

33. Chuahan NB, Siegel GJ. Reversal of amyloid β toxicity in Alzheimer's disease model Tg2576 by intraventricular anti-amyloid β antibody. J Neurosci Res 2002; 69:10-23.

34. Morley JE, Farr SA, Flood JF. Antibody to amyloid β protein alleviates impaired acquisition, retention, and memory processing in SAMP8 mice. Neurobiol Learning and Memory 2002; 78:125-138.

35. Solomon B, Koppel R, Hanan E et al. Monoclonal antibodies inhibit in vitro fibrillar aggregation of the Alzheimer β-amyloid peptide. Proc Natl Acad Sci USA 1996; 93:452-455.

36. Solomon B, Koppel R, Frenkel D et al. Disaggregation of Alzheimer β-amyloid by site-directed mAb. Proc Natl Acad Sci USA 1997; 94:4109-4112.

37. Bacskai BJ, Kajdasz ST, Christie RH et al. Imaging of amyloid-β deposits in brains of living mice permits direct observation of clearance of plaques with immunotherapy. Nat Med 2001; 7(3):369-372.
38. Bacskai B, Kajdasz ST, McLellan ME et al. Non-Fc-mediated mechanisms are involved in clearance of amyloid-β in vivo by immunotherapy. J Neurosci 2002; 22(18):7873-7878.
39. Pan W, Solomon B, Maness LM et al. Antibodies to beta-amyloid decrease the blood-to-brain transfer of beta-amyloid peptide. Exp Biol Med 2002; 227(8):609-615.
40. Schenk D. Amyloid-β immunotherapy for Alzheimer's disease: the end of the beginning. Nature Reviews Neuroscience 2002; 3:824-828.
41. Grubeck-Loebenstein B, Blasko I, Marx F et al. Immunization with β-amyloid: could T-cell activation have a harmful effect? Trends in Neurosci 2000; 23(3):114.
42. Marx F, Blasko I, Pavelka M et al. The possible role of the immune system in Alzheimer's disease. Exp Gerontol 1998; 33:871-881.
43. Marx F, Blasko I, Zisterer K et al. Transfected human B cells: a new model to study the functional and immunostimulatory consequences of APP production. Exp Gerontol 1999; 34:783-795.
44. Munch G, Robinson SR, Potential neurotoxic inflammatory responses to Abeta vaccination in humans. J Neural Transm 2002; 109:1081-1087.
45. Hock C, Konietzko U, Papssotiropoulos A et al. Generation of antibodies specific for beta-amyloid by vaccination of patients with Alzheimer's disease. Nat Med 2002; 11:1270-1275.
46. Imbimbo BP, Toxicity of β-amyloid vaccination in patients with Alzheimer's disease. Annals Neurol 2002; 51(6):794.
47. Newman MJ, Wu J-Y, Gardner BH et al. Induction of cross-reactive cytotoxic T-lymphocyte responses specific for HIV-1 gp120 using saponin adjuvant (QS-21) supplemented subunit vaccine formulations. Vaccine 1997; 15:1001-1007.
48. Kensil CR, Wu JY, Anderson CA et al. QS-21 and QS-7: purified saponin adjuvants. Dev Biol Stand 1998; 92:41-47.
49. Singh M, O'Hagan D. Advances in vaccine adjuvants. Nature Biotechnology 1999; 17:1075-1081.
50. Kenney RT, Rabinovitch NR, Pichyangkul S et al. Meeting Report: 2nd meeting on novel adjuvants currently in/close to human clinical testing, World Health Organization, France, 5-6 June 2000. Vaccine 2002; 20:2155-2163.
51. Foon KA, Sen G, Hutchins L et al. Antibody responses in melanoma patients immunized with an anti-idiotypic antibody mimicking disialoganglioside GD2. Clin Cancer Res 1998; 4(5):1117-1124.
52. Foon KA, Lutzky J, Baral RN et al. Clinical and immune responses in advanced melanoma patients immunized with an anti-idiotypic antibody mimicking disialoganglioside GD2. J Clin Oncol 2000; 18(2):376-384.
53. Evans TG, McElrath MJ, Matthews T et al. QS-21 promotes an adjuvant effect allowing for reduced antigen dose during HIV-1 envelope subunit immunizations in humans. Vaccine 2001; 19:2080-2091.
54. Gilewski T, Adluri S, Ragupathi G et al. Vaccination of high-risk breast cancer patients with mucin-1 (MUC1) keyhole limpet hemocyanin conjugate plus QS-21. Clin Cancer Res 2000; 6(5):1693-1701.
55. Poduslo JF, Curran GL. Amyloid β peptide as a vaccine for Alzheimer's disease involves receptor-mediated transport at the blood-brain barrier. NeuroReport 2001; 12(15):3197-3200.
56. Dodel R, Hampel H, Depboylu C et al. Human antibodies against amyloid β peptide: A potential treatment for Alzheimer's disease. Ann Neurol 2002; 52:253-256.
57. Brayden DJ, Templeton L, McClean S et al. Encapsulation in biodegradable microparticles enhances serum antibody response to parenterally-delivered β-amyloid in mice. Vaccine 2001; 19:4185-4193.
58. Nicolau C, Greferath R, Balaban TS et al. A liposome-based therapeutic vaccine against β-amyloid plaques on the pancreas of transgenic NORBA mice. Proc Natl Acad Sci USA 2002; 99(4):2332-2337.
59. Cao C, Bai Y, Ugen K et al. β-amyloid vaccine strategies for Alzheimer's disease (AD). Neurobiol Aging 2002; Abstr. 399:S105.
60. Ghochikyan A, Agadjanyan M, Cribbs, DH. Development of a vaccine against animal models of Alzheimer's disease. Neurobiol Aging 2002; Abstr. 888:S237.

61. Das P, Minidis N, Golde, T. Effectiveness of Aβ1-42 immunization in the Tg2576 mouse model using DNA vaccination. Neurobiol Aging 2002; Abstr. 411:S109.

62. Alexander J, del Guercio MF, Maewal A et al. Linear PADRE T helper epitope and carbohydrate B cell epitope conjugates induce specific high titer IgG antibody responses. J Immunol 2002; 164(3):1625-1633.

63. Kim SK, Ragupathi G, Musselli C et al. Comparison of the effect of different immunological adjuvants on the antibody and T-cell response to immunization with MUC1-KLH and GD3-KLH conjugate cancer vaccines. Vaccine 1999; 12:597-603.

64. Bradney CP, Sempowski GD, Liao H-X et al. Cytokines as adjuvants for the induction of anti-human immunodeficiency virus peptide immunoglobulin G (IgG) and IgA antibodies in serum and mucosal secretions after nasal immunizations. J Virol 2002; 76(2):517-524.

65. Proietti E, Bracci L, Puzelli S et al. Type I IFN as a natural adjuvant for a protective immune response: lessons from the influenza vaccine models. J Immunol 2002; 169:375-383.

66. Marinaro M, Boyaka PN, Jackson RJ et al. Use of intranasal IL-12 to target predominantly Th1 responses to nasal and Th2 responses to oral vaccines given with cholera toxin. J Immunol 1999; 162:114-121.

Index